BURLINGTON
BREWING

BURLINGTON
BREWING

A History of Craft Beer in the Queen City

JEFF S. BAKER II AND ADAM KRAKOWSKI

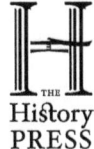

THE
History
PRESS

Published by The History Press
Charleston, SC
www.historypress.net

Copyright © 2019 by Jeff S. Baker II and Adam Krakowski
All rights reserved

Front cover: (top) The Burlington skyline. *Courtesy of David Seaver Photography.*

Back cover: (left) A fresh pint of lager at Zero Gravity Craft Brewery's Pine Street location. *Courtesy Paul Sarne*; *(right)* Steve Polewacyk pulling a beer at Vermont Pub & Brewery. Greg Noonan can be seen smiling over Polewacyk's shoulder. *Courtesy Vermont Pub & Brewery.*

First published 2019

ISBN 9781540239365

Library of Congress Control Number: 2019935345

For Noella and Margaret.

In memory of Greg Noonan.

CONTENTS

CONTENTS

FOREWORD

I n 1988, the beer scene in Burlington, Vermont, pretty much consisted of Miller, Coors, Budweiser and White River Junction upstart Catamount Brewing Company. But by the end of that year, Greg and Nancy Noonan would introduce the concept of a brewpub, a relatively new idea on the East Coast at the time.

The Noonans scoured New England looking for a suitable place, and when they arrived in Burlington, the search finally ended. Armed with a shoestring budget and some used stainless steel equipment, including a maple sap boiler, an old pig feeder and an ice cream maker, Greg fashioned a commercial brewery using pure Yankee ingenuity. But Vermont laws wouldn't allow a brewery to sell its beer for on-site consumption. New laws needed to be enacted to allow beer to be served on the premises where it was manufactured, and today, every Vermont brewery with a taproom can thank Greg Noonan and former legislator Bill Mares for getting the job done.

Thus, Vermont Pub & Brewery, Burlington's first brewpub, was born. But in the early years of the craft beer movement, local consumers were accustomed to mainstream, mass-produced yellow beer. Consumer education was needed to make this project work. The market was soon to be augmented with robust, full-bodied and hoppy beers. Each day at the new fourteen-barrel brewpub blended into the next as a new beer scene began to grow. Nearly everything was done by the seat of our pants, including the new and creative beer recipes that Greg was coming up with…in the early '90s, Brettanomyces beers were practically unheard of. In 1989, Greg helped to start the fledgling Green Mountain Mashers homebrewing club to enhance the new beer culture that Burlington would eventually become famous for with a focus on education.

While the craft beer environment was developing all over Vermont in the 1990s, Burlington was also growing. Three Needs Brewery & Taproom was founded by Glenn Walter and was soon followed by Zero Gravity Craft Brewery, led by former Catamount brewer Paul Sayler. Former Green Mountain Mashers homebrewer and winner of multiple Brew Masters Cups Paul Hale opened Queen City Brewing. Magic Hat, Switchback, Foam Brewers, House of Fermentology and Simple Roots Brewing have helped to populate the Burlington brewing landscape over the years.

Today, there is a new generation of brewers and consumers who are better educated and have more discerning palates and an enthusiasm like never before. As the Burlington beer scene continues to evolve and reaches heights never imagined in those beginning days, I am humbled to have been part of it and observed its ever-growing progress. Moving forward, I don't necessarily see an end to this growth and welcome all the new ideas inspired by today's Burlington brewers. I offer a tribute to both Greg and Nancy for their inspiration and pioneering of what has become the burgeoning craft beer institution in the Burlington community.

—Steven Polewacyk
Vermont Pub & Brewery

ACKNOWLEDGEMENTS

As ALWAYS, FIRST AND most thankful for my wife, Noella, who has fully supported me in pursuing writing. Without her, I would never accomplish what I have been able to do. Every dream. I would also like to sincerely thank our research assistant Ian Toshio Miyashiro, who assisted in transcribing historical documents as well as interviews for the production of this book. I want to thank all the brewers who opened their doors for this project and took time to help. Finally, I wish to thank my parents, Andrzej Krakowski and Majka Elczewska, as well as Ed and Elli Giobbi, for pushing and inspiring me to strive and pursue my passion. Thank you.

—Adam Krakowski

FIRST, I COULD NOT have finished this project without the constant encouragement from my partner, Jana. Her love of literature and the arts is truly inspiring. Thanks to my parents, Margaret and Joel, for all their support and willingness to tour every darn brewery. For many of the interviews for this book, I was assisted by Ryan "the Can Stealer" Chaffin and Jason "J Rock" Strempek. Thank you to David J. Farrell, Jed Davis, Paul Sayler, Jennifer Swiatek, George Bergin and Todd Delbeck for employing me in the Vermont beverage industry and for coaching me along the way. Props to Ian Toshio Miyashiro and Noella Krakowski for their technical and editing assistance on the manuscript. Cheers to my great friend and fellow nihilist, Paul Sarne, who provided his professional eye for a number of photographs in this book. And special thanks to Steve Polewacyk, a truly rare character with whom I greatly enjoy drinking beer.

—Jeff S. Baker II

PREFACE

Over the last decade or so, Burlington—Vermont's largest city—has emerged as a culinary and craft beer destination thanks to a renaissance of brewing and a burgeoning farm-to-table restaurant scene. Numerous Burlington breweries and restaurants have graced the pages of leading periodicals, but those who've never been to Burlington might ask: What makes Vermont's Queen City so special? Well, simply put, the Burlington scene has garnered its national—nay, international—reputation by being entirely homegrown and developing organically independent of widespread trends. (Okay, maybe Vermont started the kale trend, but we promise that the wheatgrass shot was not from up here!)

Starting from its modern brewing era in the late 1980s, Burlington has been the epicenter of Vermont's brewing scene. From the godfather of Vermont beer, Greg Noonan, to the next brewery that will open its doors in the city, Burlington is at the forefront of national brewing discussions. When compared to the Irish traditions in Boston or the German traditions in Milwaukee, there was no real brewing tradition to speak of in Burlington, so it has never been bound by traditional parameters. This afforded early brewers like Noonan, Bill Cherry and Paul Sayler to curate a wide array of styles and share their own interpretations. Sometimes, their beers have been bracingly true to the original European styles, while other times they have been completely unique. This allowed for a distinctive brewing scene to emerge compared to other beer-soaked places around the country.

Before taking a deep dive into the brewing history and current beer culture in Burlington, think for a moment about what season it is. This may seem like an awkward way to begin the book, but stay with us for a moment. Seasonality has faded from modern mainstream culture. Sure, we all know the different seasons (in Vermont, we claim six seasons, adding "Stick Season" and "Mud Season"), but our food heritage and traditions are much more unique than we realize; we no longer fully understand seasons in terms of cooking and brewing. In today's culinary landscape, we all have seasonal favorites, such as ramps (wild leeks), fiddlehead ferns, and morel mushrooms in the spring, for example, but we also eat a lot of things out of season. Take tomatoes, for example; come winter, the tomatoes available at the grocery store are pink in color and flavorless, yet people still purchase and eat them.

Yet, as recently as seventy years ago, home preservation was commonplace, while it is now more of a choice than a necessity. Most of the pork and beef that was consumed one hundred years ago came from the brine barrel or had been cured, smoked or dried to keep throughout the winter. Fresh meat had a short shelf life and was a truly special treat. It was much the same with brewing.

Breweries in nineteenth-century Vermont were not year-round operations as they are today, as brewing was limited—mainly due to resources. When a brewery's grain stocks were spent, that concluded the brewing season. There are numerous examples in the *Burlington Free Press* and other newspapers from the 1830s and 1840s offering to pay cash for barley delivered to the brewery. Often, a notice or advertisement was placed in the papers, usually in the first weeks of September, just after both the grain and hop harvests became available. The brewing season would extend through April or May at the latest. While the breweries could possibly still have finished ale resting in casks, there was no more barley nor hops until the next harvest. Larger breweries from Boston; Troy and Albany, New York; Portsmouth, New Hampshire; and the Province of Quebec in Canada were able to fill the voids by sending beer in barrels and bottles by boat or on horseback and, later, by rail and automobile.

You may not realize it, but most of the beers we drink today are "out of season." Historically, certain styles were only brewed at specific times of the year. A Maibock, a strong and dark German lager, was brewed to fortify the labor for the spring planting, while Märzens were lagered over the summer for Oktoberfest celebrations in the fall.

Speaking of fall, pumpkin beers have, over the last decade, been emerging as early as July—well before the northern hemisphere's pumpkin harvest. Utilizing pumpkin in the nineteenth century was done out of necessity, not to capture the taste of pumpkin pie. When the grain harvest was weak or stores ran low, pumpkins were added to the mash to boost fermentable sugars. And spices were never added, since, at the time, they were some of the most valuable pantry items. Cinnamon, cloves and ginger all came from the Far East or West Indies and were quite costly. They were not tossed into beer.

When hops were in short supply or the crop was poor, native herbs and botanicals such as mugwort, sage, bog myrtle and sweet gale were substituted as a bittering agent to balance the sweetness of the malt. This style has been known through history as gruit and, again, was far more of a necessity in the nineteenth century than merely a pleasurable venture. Burlington's Zero Gravity Craft Brewing would be the first to revive the tradition with locally grown in-season herbs and botanicals.

While this discussion of seasonal brewing might seem trivial, it is an example of how—in this day and age—people are able to erase seasonality. The brewing and culinary worlds have changed over the last three decades to produce *for* a season rather than *from* a season. Historically, seasons were a rhythm but also a boundary.

Today, a serious volume of beer flows from the city of Burlington. The breweries are open year-round, and the brews are in very high demand. The history of beer in Burlington—a city that went for one hundred years without a brewery and now has ten breweries, a distillery and a craft cidery—is rich with rare characters, fascinating stories, vintage bars, notable architecture and tales of hauntings. A confluence of passion, camaraderie and a dedication to the craft connects all of these brewers working long hours. They don't do it for the praise, but thankfully, the rest of the country has noticed what makes the Queen City so special.

PART I
HISTORY

THE EARLY DAYS

W hen it comes to making beer, the core concept of brewing has remained the same since its discovery. The alchemy of ingredients used today (grains, hops, water and yeast) are the same as those implemented in the sixteenth century with the German purity law known as the Reinheitsgebot. Two centuries ago, the role of the brewer was much different. The skills a brewmaster needed to operate a brewery were different. The brewer had to be trained in all aspects of making beer: brewing, producing and procuring the ingredients.

The modern process of making beer has not changed, although brewers have access to more efficient and advanced equipment. It has pushed many boundaries and utilized new ingredients, but the core of brewing has remained the same. The transportation methods for moving grains, hops and anything else in the world have allowed for expedited shipment of ingredients. This has helped to streamline the focus of the brewer to hone a very high precision craft.

The main separation in abilities of the brewers of now and those of centuries ago comes down to malting. The process of malting grains allows natural enzymes in the grain to convert the complex carbohydrates stored within the grain to simpler sugars that can then be fermented with hops and yeast into beer. Malting, though, is very hard work. Only recently, the art of malting has returned to the Green Mountains with Peterson Quality Malt in Monkton, Vermont. A profession that was once a central part of every town barely exists statewide.

In late seventeenth-century Vermont, malthouses were an important pillar of communities. Malthouses served the entire community and allowed the farmers in the area to create value-added product in malted barley as well as the milling of the grains. Often, the maltster and miller would have an agreement where they got to keep a portion of the harvest they were processing. Breweries and distilleries were often side operations attached to early malthouses in Vermont. Some went further and produced pot and pearl ashes from the by-product of the kilns used for the malting process.

In 1800, the burgeoning city of Burlington had its first commercial brewery. Daniel Staniford's joint brewery and distillery was located near the northeast corner of Pearl Street and present-day Winooski Avenue.[1] The positioning of the brewery—near the newly founded University of Vermont—was not without warrant. A brewery close, if not at, the university was an asset for students. In the current context of society, we look at alcohol as a social lubricant. At the turn of the nineteenth century, picture something else. At that time, well before germ theory and Louis Pasteur, beer was a form of nutrition in the source of stored carbohydrates and vitamins. The yeast is rich in vitamin B-complex and trace minerals. This source of nutrition was valuable to a university and its students. The idea of having a brewery in close proximity to the university was a tool for higher education. It meant that in an age of seasonal food, the students did not have to worry and could focus on their studies.

Staniford's brewery produced "beer, ale, and porter, and manufactured other fluids which even the phlegmatic votary of lager cannot claim as non-intoxicating."[2] Staniford took out an advertisement in the *Vermont Cenitnel* on March 14, 1803. It promoted his gin and also informed the public that due to scarcity of barley from that season, he would not be producing his Strong Beer that spring. He made his intentions known that he shall be brewing his beer, ale and porter as soon as the barley was available that season. The advertisement also included a classified ad looking for a young boy around fifteen years of age to come on as an apprentice and a "loyal" journeyman to assist in the operation.

By 1805, Staniford's brewing operation had expanded. He now had a brewer, Philip Castle, who was offering up both strong and table beer at the distillery/brewery. Before coming to Burlington, Castle had brewed in Dublin, Ireland, for two years and in London, England, as an apprentice.[3] A table beer was made by taking the second runnings of a stronger beer's mash and using it for wort. Staniford had brewed a strong beer at his operation, but Castle's addition of a table beer was something new.

Breweries at the turn of the eighteenth century were often run in conjunction with distilleries, such as at Staniford's operation. This was most likely due to having a malting floor to convert the raw grains into fermentable ones. Both operations use the same core processes and ingredients. It is easy to see why early brewers in Vermont were producers of both ales and spirits. Daniel Staniford's operation, as well as his property, was announced for sale on March 24, 1831.[4] In the advertisement, the property listed was the distillery operation. This gives the distillery around thirty years in operation. Looking at the total amount of property and businesses included in the listing, Staniford was very successful in his endeavors.

Shortly after Staniford started his brewing and distilling operations in Burlington, farther down, at the head of Pearl Street, another operation by Loomis and Bradley emerged with a similar operation. It is not known how long the operation lasted. We know the operation started shortly after Staniford's. By the time of the 1830 state census, the operation by Loomis and Bradley is no longer listed. The last advertisement placed by Loomis & Bradley, inquiring about rye grain for sale, was on September 7, 1827, in the *Burlington Free Press*. The ad stated that Loomis & Bradley, who also operated a dry goods store in Burlington, would be willing to pay cash or half goods for the procurement of rye grains delivered to the distillery. It is important to note that there were further advertisements present in the papers through 1827 and 1828. These requested corn, barley, rye and sugar. They make no mention of the distillery, instead requesting delivery to the store. The brewery is believed to have been part of the distillery.

Another goods store owner who started a brewing operation in Burlington was Samuel Hickok. He started his operation on the west side of Champlain Street. Hickok placed an advertisement in the *Burlington Free Press* on March 20, 1828, expressing his intention to open a brewery. He also wished to purchase 6,000 bushels of barley (equivalent to around 270,000 pounds), along with 200 to 300 pounds of seed barley. By October of the same year, an announcement in the *Burlington Free Press* stated that the Burlington Brewery had fresh ale for sale:

BREWERY.
THE undersigned, intending to establish a Brewery in this place, will pay cash for
6,000 bushels of
BARLEY,
delivered next fall and winter.—He also wishes to purchase 2 or 300 bushels for SEED, delivered between this and the 1st of May next, for which a liberal price will be paid on delivery.

Provided the whole or any part of said seed shall be obtained, it will be offered for sale to those who may think it an object to commence the growing of this article.
SAMUEL HICKOK.
Burlington, March 5, 1828.[5]

A month later, an advertisement had the title of "Burlington Ale," which was for sale at the brewery and kept in constant supply. According to the advertisements and newspapers, Hickok's Burlington Brewery only produced Burlington Ale—there was no mention of any other type of ale being produced. Advertisements were present for barley or other grains, but it seemed there was never an issue of procuring enough hops for the brewing operation, as there are no requests or mentions of hops in the same advertisements printed in and around Burlington. The brewery's advertisements for having ale on hand appeared from late September through the early spring of each year. There were no advertisements for the brewery running in papers during the late spring and summer months. This suggests that the brewery was seasonally dependent on weather conditions and supplies of barley. Every time an advertisement ran for the purchase of barley, the advertisement stated that deliveries run through the fall and winter. And every year, the design of Hickok's Burlington Ale advertisement changed:

BURLINGTON BREWERY.
These works, which have been erecting in this village during the past season, by Mr. E. Hickok, for the purpose of furnishing the citizens of Vermont a healthful malt liquor in the place of foreign distilled spirits, are now completed, and the establishment is in the full tide of successful operation. The Beer already produced is pronounced by good judges to be equal, if not superior to that imported here from Lansingburgh and Albany. Here is another home market for the hops and barley of the farmer, and the large sums which have heretofore been sent abroad for this article which has already come into so general use, will hereafter be kept for home circulation.[6]

While Burlington Ale was prominently advertised, no other goods store in the city made any mention of ale for sale. The first advertisement featuring imported ale in Burlington is from a January 1832 advertisement in the *Burlington Free Press*. The advertisement announces the opening of a wet and dry goods store by Jonathan Allen on the corner of Church and College Streets. The ad listed all of the available products for sale. Among the extensive list are barrels of "Ale & Philadelphia Porter."[7] This advertisement for Philadelphia Porter shows that importation of beer from a far distance was possible. This led to the possibility that ales from Boston, Albany or Portsmouth—and likely from points farther away—were in the Vermont marketplace.

The Hickoks sold the brewery in October 1832 with Hickok's goods store. An advertisement placed in the *Burlington Free Press* on October 12, 1832, stated that the stock of the brewery as well as the brewing operation itself has been purchased by H.W. Catlin & Co. It is important to note that the last lines of the advertisement state that cash will be paid for any quantity of good quality barley and twelve or thirteen hundredweight (cwt) of well-pressed hops. This is the earliest known advertisement for hops in Vermont. The request for 1,200 to 1,300 pounds of hops suggests that such large amounts of hops were available in the Chittenden County region. With new ownership, the brewery was still operational. It still produced Burlington Ale; no other recipe was produced. A noticeable change is that the Burlington Brewery no longer had its own individual advertisement for its wares but was reduced to a line: "Burlington Ale is now operational, orders for ale will be promptly attended to."[8] On September 12, 1834, Catlin & Co. was dissolved. It became Hickok & Catlin, with H.P. Hickok listed as a partner. The brewery was still listed as the Burlington Brewery. Now, there was a listing of not only full barrels but half barrels for sale.[9] This was possibly due to Catlin & Co. placing an advertisement the year before for a journeyman cooper and a need for ten thousand white oak staves of barrel quality.[10] At some point in the winter of 1835–36, the Burlington Brewery was sold by Hickok & Catlin to Robert N. Flack, a Lake Champlain sea captain and Irish immigrant.[11] Flack retained the name Burlington Brewery but no longer produced Burlington Ale—at least not by that name. By the fall of 1836, Flack was producing "Flack's No. 1" as well as one to two other ales that were lauded in a *Burlington Weekly Free Press* article and were in high demand.[12]

BEER —The Bostonians are gambling at no small rate about their beer. The brewers, they say, are drugging them with vile concoctions of aloe, liquorice, capscum, and John Barleycorn knows what not—one draught of which is considered a full portion for an adult. What a pity our rail-road is not completed. A few pipes of FLACK's No. 1 would give these Bostonians new conceptions of the superior virtues of good malt and hops, steeped in not too much pure water; and we doubt not they would very soon learn to "repeat the dose" as often as the Scotch parson's rule of moderation would allow. Probably there is no better article of the kind manufactured in the United States than is now furnished at the Burlington Brewery—a fact already well appreciated, if we may judge from the extended demand for it.—So long as beer is used at all, let us encourage the genuine article, and go for those who stick to malt and hops, and a plenty of them.[13]

The brewery continued to advertise for the sale of ale and the purchase of barley. Little more is known of the operation other than one advertisement placed in 1837. The advertisement from R.N. Flack was for the return of his strayed eight-year-old red-and-white spotted cow to the Burlington Brewery with the finder "handsomely rewarded."[14] Flack operated the brewery until July 6, 1839, when a fire of criminal origin decimated the building.

ANOTHER FIRE.—A fire broke out between two and three o'clock yesterday morning, in the Brewery in this village, owned and occupied by R.N. Flack, Esq. which in a short time was entirely consumed, together with a large building adjoining, belonging to the same establishment. The engines were promptly on the ground, and rendered essential service in preserving several buildings adjacent. There is no doubt that his fire was the work of an incendiary, as we understand one attempt previous to this has been made to fire the same premises within the last fortnight. This is the fourth successful act of incendiarism which has occurred in this village within a year, and which has resulted in a large destruction of property. Can no means be devised to ferret out the scoundrels?

We learn that there was an insurance of $4,500 on the buildings and stock, $3,000 in the Mutual, and $1,500 in the Ætna company.[15]

Flack had insurance on the building, but it was difficult to resurrect the brewery. The Burlington Brewery was one of an estimated six or seven buildings burned by arson over the course of 1829 in Burlington. It would be easy to blame the arson on a radical religious or temperance movement individual. But other buildings struck by the fires included churches and other structures. Flack sold the property and company name to another Burlington resident. The Burlington Brewery returned to operation under the ownership of George Peterson on September 11, 1840.[16] During the first year of operation, Peterson advertised that he had ale on hand into the summer:

BEER.

The Burlington Brewery having been rebuilt during the past season, will be put in operation on Tuesday next, and will in a few days, be prepared to answer orders for BEER, to any extent. The apparatus is all new, and of the most approved construction; and it a first rate brewer-good materials, and a firm resolve on the part of the proprietor to furnish a superior article of beer, can avail any thing, my customers shall find no cause of complaint. The public are respectfully invited to give the new establishment a fair trial. GEO. PETERSON Burlington, Sept. 20 1839[17]

The brewery was producing around 1,500 barrels of ale per year.[18] By the time of the 1850 Vermont agricultural census, Peterson's operation was only two men producing 500 barrels of ale. This had a value of $3,000, which equaled $6 for a barrel of Burlington Brewery beer. The most telling information on the census was the information Peterson provided. To produce the 500 barrels of ale, Peterson used 2,500 bushels of barley along with 2,000 pounds of hops. The Burlington Brewery continued operations past the onset of the 1852 prohibition law in the state of Vermont, although it had a far less public image, removing its advertisements from the newspapers.

With the onset of state prohibition in 1852, the Burlington Brewery remained out of the news, only appearing in notices regarding legal issues. It simply brewed its ales, then promptly sent them over to Plattsburgh for sales and distribution. After keeping a low profile for fifteen years, the next entry of note for the brewer was in a January 11, 1867 article from the *Burlington Free Press*. It recorded that forty-seven dollars were stolen from the brewery. The next time the brewery made the news, it was far more significant. Monday, June 21, 1867, marked the start of the terminal blow that would put an end to brewing in Vermont for a century.

That afternoon in Burlington, Sheriff Munson and Chief of Police Drew, accompanied by officers, executed search warrants for illegal liquor around the city. One of their search-and-seizures in Burlington was at Peterson's brewery. Officers seized one hogshead, four barrels, one half-barrel and four quarter-barrels of ale. There was an added barrels' worth of bottled ale. The seizure of ale and subsequent arrest of George Peterson's son, Benjamin, the owner and brewer of Burlington Brewery, was significant. It marked the first time under a state or federal prohibition that a legal and established brewer of beer was arrested for the nature of his work. Peterson's case was heard before the Chittenden County court on August 2, 1867. It went through the Vermont legal system, culminating with his case being heard during the January 1869 term of the Supreme Court of Vermont.

The details of the case before the Vermont Supreme Court offer an important insight into both Peterson's operation and the cultural mindset of Vermont. The key witness in the prosecution's case was William Davis. Davis was a teamster employed by Peterson to carry away the beer. During the course of his testimony, he divulged that most of Peterson's beer "goes across the lake [Champlain]." Peterson was accused of producing "strong beer." In the eyes of Vermont, it was seen as exceeding 3.2 percent alcohol by weight, so it was no surprise that Peterson shipped the beer out of state. At the time Peterson's brewery was in operation, Plattsburgh, New York, was a major distribution hub for goods. It is possible that the flagship beverage of the brewery, Burlington Ale, was primarily produced and shipped out of town. It is further possible that Peterson used the second runnings of the malt; this would produce a small beer. The small beer would have an alcohol level that fell below the threshold, allowing for local sales.

The second witness called by the prosecution was Luman A. Drew, a curious character in the case. Drew testified that strong beer was intoxicating, though Drew provided no credentials to support his authority. He further stated that strong beer was produced at Peterson's brewery. He also admitted that he never personally witnessed the strong beer being produced. Under cross-examination, the witness admitted that it was only hearsay; he never saw Peterson producing strong beer at all.

The next witness was Thomas Rhodes. After being told by the judge that his testimony would not be self-incriminating, he testified that roughly nine months prior, he purchased a half-barrel of strong beer from Peterson. It was delivered to the Rutland & Burlington rail depot.

The final witness called by the prosecution was Edgar Burritt. Burritt was a druggist based in Burlington. While on the witness stand, he was shown a bottle of beer from Peterson's brewery. He was further asked what the liquid was. Burritt surprisingly informed the court that it was a sour hop beer. It was either strong beer or small beer, and he did not know anything further about the liquid. His testimony concluded the prosecution's case.

Peterson was ultimately found guilty of the original charge and subsequently fined. Benjamin Peterson sold the brewery to John W. Carpenter on July 24, 1872. At the time, it was noted that the brewery was "in the business of bottling cider, soda, and mineral waters."[19] Due to issues with the sale, a court case ensued. One of the enticing details in the case was that ale, lager and porter were found on hand. By the time of the sale, the beer had been "soured" and was unfit for consumption unless it were to be blended. Presumably, Peterson was still making his Burlington Ale. The important takeaway is that Peterson had porter and lager on hand. No porter or lager was listed in the seizure at his brewery. It is possible that he started producing or had been producing these in addition to his Burlington Ale.

Once the difficulties with the law arose, the sale of the Burlington brewery fell through. Peterson continued to try to sell the brewery under the rule of state prohibition. He found a buyer; Ammi F. Stone, a retired lawyer and merchant from Charlotte, took over the Burlington Brewery with his son William. They carried on the operation for nearly a decade, brewing around 3,000 barrels (100,000 gallons) until 1879. The Burlington brewery started as an operation of brewing and bottling lager in 1871. The brewing operation moved to Albany, New York, "on account to the stringency of prohibition law" sometime in 1882 or 1883.[20] The record in the August 2, 1882 issue of the *Argus and Patriot* notes fifteen bottles of beer seized at the brewery (Stone's) and forfeited. The proprietor (Stone) was found guilty and fined for the infraction. This specific seizure is fascinating. The bottles contained beer. But, unlike in previous seizures, there were no notations about the contents. All previous seizures documented the contents as ale, strong beer, lager or porter. Aside from that, William Stone turned the brewery into a successful bottling company, where the bottling of mineral and soda water carried on into the twentieth century.

There was one last attempt to start a brewery in Burlington in 1906, before the onset of federal Prohibition. A petition (H. 57) came to the legislation to once again create the Burlington Brewery Company. On November 15, 1906, it was effectively stopped—or rather, suffocated. It was a classic case of not wanting to have names attached to a very delicate subject. There was an attempt to remove the petition from the committee on corporations, then a debate to first move it to the committee on public health. Then, after being voted down, it went to the judiciary committee, with the same result.[21]

By 1850, momentum for the temperance movement in Vermont had been building for well over a decade. However, the state was lacking a catalyst to push the temperance efforts from discussion into law. The spark that ignited the "building powder keg" of policy in the state came from Maine. The 1851 Prohibition Law in Maine, spearheaded by the efforts of Neal Dow, the mayor of Portland, was passed. The "Maine Law," as it was referred to in Vermont newspapers, was the first law of its type that prohibited the manufacture and sale of liquor statewide in New England. Over the years, it led to political and social unrest in Maine. This culminated in riots such as the Portland Rum Riot of 1855, which was over medicinal liquor held at city hall. A swelling in crowd numbers resulted in the militia being called and violence ensuing. The effects of the riot led Maine to repeal the state's prohibition law in 1856.

Before the Maine Law, Vermont started to experiment with prohibition, starting at the county level rather than the state level. The General Assembly of Vermont passed the first law that was a precursor to state prohibition on November 3, 1846. Known as Act 24, it targeted the operation licensing of innkeepers and retailers of spirits in the state on a county-by-county vote. It allowed a statewide vote each March, beginning in 1847, to decide whether to allow licenses to sell alcoholic beverages for that year. The first vote, in 1847, resulted in a 21,798 to 13,707 vote to ban licenses in nearly all counties. Only Essex County voted to keep issuing licenses for the sale of alcohol.[22] This vote left Vermont a "dry" state (with the exception of Essex County). The ban was, by many accounts, a symbolic one and not readily enforced. The vote of 1848 changed course regarding the bans on licensing, as a majority of counties favored licensing. The temperance vote succeeded again in 1849 and 1850 to keep Vermont dry; Essex County also voted to be dry in those years.

The passing of the Maine Law strengthened attitudes at both the political and social levels in Vermont. A newspaper later remarked that "the operations of this law were watched with eager interest by a large number of Vermonters who were dissatisfied with the legal methods provided in the state for dealing with the drink evil, and the new system soon found many enthusiastic advocates."[23]

The state elections that year did not give a clear majority to any gubernatorial candidate. The Whig Party failed to address the issue of slavery. This allowed for a rapid gain in the strength of the Democratic Party and pulled from the ranks of the Whig Party. The erosion of the once powerful Whig Party became evident when the votes were spread out between candidates. On October 14, 1852, the Vermont legislature convened. The twenty-fourth ballot elected Erastus Fairbanks from St. Johnsbury as governor. Thomas Powers of Woodstock was selected as speaker of the house. William Kittredge of Fair Haven became lieutenant governor. Fairbanks, a Whig and a staunch supporter of prohibition, was now in a position to push temperance legislation further.

A committee was hastily created to analyze the Maine Law and create one that would fit Vermont. Some members appointed to the committee were not keen on serving. Mr. Sherman of Montgomery requested to be removed from the committee. His view "was not so sanguine as some others that the Maine liquor law would answer the purposed designed." Representative Cushman of Fair Haven also chose not to serve on the committee. Two weeks later, the committee's work was introduced by Free Democrat General Horatio Needham of Bristol. The bill was very similar to Maine's liquor law, but there were some different and important changes. It was ordered that the bill proceed and that five hundred copies of the bill be printed for review. During the course of the committee's work on crafting a bill for prohibition, petitions were flooding in from all corners of the state. A total of 38,000 signatures arrived in Montpelier, 17,500 of which belonged to legal voters, undoubtedly giving immense support to the committee's efforts.[24]

While the bills were being considered, the push for prohibition got further support. Neal Dow, the "apostle of prohibition," came to Montpelier on November 9. He was there to advise the legislature on temperance legislation. A week later, on November 17, amendments were presented for a bill titled "to prevent the traffic in intoxicating liquor." One amendment introduced would allow the manufacturing, sale and use of "the fruit of the vine" for religious purposes. Another mandated that each town elect a liquor commissioner. Other amendments introduced were voted down in the senate.

That Saturday, November 20, support surged behind the law. In a series of very quick motions that day, the bill was brought for consideration on additional amendments. After consideration of the amendments, the bill was ordered for a third reading before the senate by a vote of twenty-two to six in favor. The bill was passed the same day by a vote of nineteen to six, with a few senators absent from the vote. Passing the bill in the senate led Representative Porter of Cornwall to recommend the house abandon their bill and adopt the bill from the senate. Porter stated, "The bill was necessary, constitutional, and would be effective." In opposition was Dr. Steven of Guliford, a Whig from the same party as Porter, who asserted that a majority of the people of the state of Vermont had not asked for the law and that the existing law from 1850 was sufficient in its intent. He also noted that constitutionality of the Maine Law was highly questionable and the hasty process had not looked at the bill sufficiently. Undeterred, a vote by the house saw the senate bill come up for special order on Monday by a vote of eighty-nine to eighty-four.[25]

Monday, November 22, was a turning point in Vermont history. A packed gallery in the house chamber saw the original house bill dismissed by a close vote of ninety-two to eighty-eight. The senate-passed bill was then introduced to the house around 11:00 a.m. The bill had a referendum clause attached to it. If a majority of the referendum voters were against the bill, it would not go into effect until December 1853. Mr. Barlow of Fairfield made a motion to postpone consideration of the bill until that Wednesday at 7:00 a.m. This essentially would have left the bill dead on the table, as it was likely that the session would come to an end on Tuesday. The motion was voted on and lost by a vote of ninety-four to eighty-eight.

After settling that the bill was to proceed, house members proposed a quick flurry of additional amendments. None passed a vote. One important change that did occur was the removal of a passage that would not hold up under constitutional scrutiny; the passage stated that if one were to appeal their initial conviction under the law and be unsuccessful, their fines would be doubled. The bill then came up for a third reading, after which it was immediately voted upon and adopted—with a vote of ninety-one to ninety—accompanied by loud applause from the gallery of observers. Shortly after the bill passed the house, it was quickly moved to the senate, where the house amendments were adopted without issue. The bill was then swiftly taken to Governor Fairbanks for signing.

The quick succession of events in the chambers led house member Mr. Hager of Halifax to move to reconsider the vote that evening. He and another member of the house believed they had voted incorrectly. This would change the outcome of the bill. The speaker of the house, Thomas Powers of Woodstock, informed the house that the bill was no longer in the possession of the house and could not be reconsidered. A move by Mr. Barlow of Fairfield to return the bill to the house from the senate passed with a vote of ninety-three to eighty-two. The vote came up in the senate to return the bill to the house. The chair ruled that the bill could not be returned unless all the amendments were also reconsidered. The move was made to return the bill and was overturned. Before the house had time to further pursue the issue, the senate moved to put the bill before Governor Fairbanks. Fairbanks signed the bill around 1:00 a.m. on Tuesday morning. In doing so, he brought about the onset of one of the most difficult and complicated periods in the state's history.[26]

Part of the crafting of the bill was to have it presented to voters in a unique way. The referendum presented the law to voters not as a choice of whether the bill should be upheld as law but rather which date the law should go into effect. The question of when the bill should go into effect went to voters on the second Tuesday of February 1853. The two dates for consideration were either the second Tuesday of March (on a yes vote) or the first Monday of December 1853 (with a vote of no). The results of the statewide "referendum" resulted in a vote count of 22,315 to 21,794 to enact the earlier date as the starting date of prohibition. The vote fell along a geographic split, with all western counties favoring the earlier date. On the other side of the mountains, all of the eastern counties, except Caledonia, approved the later date. Once the referendum passed, the House of Representatives debated whether or not to repeal the law.[27] The presentation of dates as a referendum rather than the law was a step taken to avert any constitutional scrutiny.

The law, referred to as the "Liquor Law," that gripped Vermont was far from perfect. It was not as encompassing as the much later federal Prohibition that began in 1920. The Vermont law had many exceptions to the overall rule. The 1852 Liquor Law was not a full prohibition on alcohol. It was created to combat distilled spirits. The law was much less strict on natural fermenting liquors. Vermont's law allowed for wine with a religious exception. "Nothing in this chapter [94] shall be construed to prevent the manufacture, sale, and use of the fruit of the vine for the commemoration of the Lord's Supper." The law also had a section that took care not "to prevent the manufacture, sale, and use of cider, nor the manufacture by any one [sic], for his own consumption and use, of any fermented liquor." The clause that followed took care to ensure "that no person shall sell or furnish cider or fermented liquor at or in any victualing house, tavern, grocery shop, or cellar, or other place of public resort." A penalty was also mandated that no person shall "sell or furnish cider or any fermented liquor to an habitual drunkard under any circumstances." This carried a ten-dollar fine that was to be paid directly to the state treasurer. Another clause in the law stated that the consumption of alcohol in one's own home was permitted as long as it did not lead to intoxication of anyone.[28]

Another key provision in the law concerned cider. The production of cider was permitted as long as it was unadulterated. This meant that no additional fermentable sugars were added. The reasoning behind it being permitted came down to two factors. The first was how simple it was to create hard cider. A simple pressing of the apples for juice would start the fermenting process in a matter of days thanks to the natural wild yeasts present on the apples. With the abundance of apple trees and orchards across the state, it would have been impossible to outlaw the production of cider. The second factor was that cider was still an important staple of daily nutrition. Water was still a dangerous gamble to consume. Beer produced at a brewery was illegal under the law. The 1852 law, whether accidentally or intentionally, likely dramatically increased cider consumption in the state. It was not until around 1880 that the production and consumption of hard cider was outlawed in the state.

The remaining part of the 1852 Liquor Law contained the framework for what was and was not permitted for production and consumption, the penalties and the due course of how to handle infractions of the law. The law constantly morphed over subsequent decades, but it did create a fairly sound foundation for officials. There were allowances for medicinal and industrial alcohol use, and there was the previously discussed home use provision. Also included was a section addressing adulterated liquors that carried an increased fine. A very arbitrary section of the law gave an officer the power to enforce the law dependent on if they found the liquor to be "intoxicating" or not. Curiously, in the 1850s and 1860s, hard cider was often found to be non-intoxicating. Beer or spirits were obviously found to be intoxicating by the framing of the law. In the coming decades, this would lead to many disputes.

Statewide, there were mixed feelings about the law. Religious and temperance groups rejoiced that a liquor law was now in place and entrenched into the social fabric. Opponents of the law expressed displeasure—not about the law itself but rather the enforcement of such a wide-reaching law across the state. They also worried about the economic implications for towns and cities. A political observer wrote, "among all the acts passed by the Legislature of Vermont, since its existence as a State, there is probably no one more likely to excite surprise and regret in considerate minds, than that recently enacted, called the liquor law."[29] Some even argued the constitutionality of the law itself. Suffice it to say, the law caused quite a commotion.

The law nearly suffered a devastating blow before the referendum vote approached in February 1853 in Vermont. The *Vermont Watchman and State Journal* in Montpelier published news on February 3, just days before the state vote, about what had recently happened in Providence, Rhode Island. Justice Curtis, in the U.S. Circuit Court, found the Maine law—the very law that the Vermont lawmakers closely followed—unconstitutional. The ruling of Justice Curtis, read by Judge Pitman in Providence, was complicated, since the case that it was referring to, *William H. Green v. Nathan M. Briggs et al.*, involved a New York citizen (Green) sending alcohol into the state of Rhode Island. At the same time Vermont's General Assembly drafted its law, the Rhode Island General Assembly adopted the Maine law as its own liquor law. It is important to note that Vermont's law mimicked

Maine's but was created by the state legislature. Rhode Island copied and passed the Maine law verbatim.

Such a headline in the papers before a vote would surely have rattled the politicians who crafted the law, which was going before the people of Vermont in only a week. The *Watchman* article published a lengthy analysis of the case and ruling and how it would not be the same case in Vermont. A key issue at the heart of the matter was that Vermont's state constitution addressed issues that hindered the Maine Law's legality in Rhode Island. The heart of Justice Curtis's ruling was that the Maine Law hindered the right to appeal; that was not an issue in Vermont's law. In fact, the provision removed from Vermont's law was the one that doubled the fine if someone lost their appeal. This was the same fault in Rhode Island. Nevertheless, the headlines did not prevent Vermont's Liquor Law from being enacted and enforced.

1200 Gallons of Beer Seized in Burlington.

The largest liqour seizure ever made in Burlington was that of Saturday morning, when Sheriff J. A. Brodie searched a freight car in the Rutland railroad yards and the bottling works of Eugene F. Fowler and secured 55 half barrels, 36 quarter barrels and five cases of beer.

Taken all together there were nearly 1200 gallons of beer seized. It took two big drays to haul the stuff from the railroad yards to the jail, where it was put in the basement for safe keeping. One of the loads was so large that four horses were required.[30]

1902

One of the most important turning points in Vermont's political, social and economic history was the gubernatorial election of 1902. The prohibition law had been on the books for fifty years. At this point, it had been suffocating the state's economy. Following a fifty-year absence of taxes paid on the production and sale of alcohol (and the auxiliary taxes on inns and taverns), calls for the end of prohibition were growing by the end of the nineteenth century. Most argue that Vermont was always a bucolic agrarian state. This was, in part, due to the economic hardship caused by prohibition. Vermont's long trial of prohibition was a serious factor in shaping the Vermont landscape of today.

The half-century of prohibition created a favorable political atmosphere that allowed the Republican Party to maintain power. Until the 1902 election, there was little resistance to the Republican political machine in Vermont. Different outside groups attempted—in vain—to launch an offensive against the incumbent rulers, but no other party was ever remotely successful. After dominating Vermont politics for more than one hundred years, the 1902 election proved to be a difficult test that altered the course of the Republican Party in Vermont.

The last decades of the 1800s saw a rapid growth in private investment in business and industry throughout the United States. Vermont, however, was still heavily reliant on agriculture. This period of industrialization never matured in Vermont. The rapid growth that occurred in neighboring states did not happen with a lack of state funding to assist with private investment—not that there was a lack of wealth in Vermont. Many residents made their fortune in the railroad, trade and other industries. The lack of state funds for investment was not overlooked in the 1902 gubernatorial election, and the blame was squarely placed on prohibition.

In the present day, it is common to have governors run many times for reelection to the office. The late nineteenth century was a different story. It was common for a governor to serve a single two-year term and leave office, often returning to business endeavors or practices. Governors returning to the private sector after serving often found many new business opportunities available to them. In 1902, the Republican Party nominated General John G. McCullough, the president pro tempore of the senate, as its gubernatorial candidate. Another member of the Republican Party vying to be the gubernatorial candidate was Percival Clement, a state senator from Rutland County, although he was quickly passed over. This action taken by the party created a fracture. No one in state government could have foreseen the repercussions. The Republican Party officially endorsed McCullough for the 1902 gubernatorial race. With no other candidate of any concern to the incumbent party, it was thought that a smooth succession would occur. For those opposed to prohibition, things were about to get interesting.

John G. McCullough's ascent in Vermont politics was swift. Born on September 16, 1835, in Newark, Delaware, he graduated from law school at the University of Pennsylvania and was admitted to the Pennsylvania bar. McCullough's stay in Pennsylvania was short. He relocated to California, where he rose to the rank of attorney general in 1863. There, he had an unsuccessful bid for reelection. He later moved to Bennington, Vermont, where he had many business interests both in and out of the state. He resided in a home that was in the family of his wife, Eliza. Today, it is known as the Park-McCullough Historic House. He was elected to the Vermont senate in 1898 and voted in as president pro tempore of the senate. As his aspirations for governor became clear, his residency status would be questioned in many newspaper articles. His many business dealings in New York prompted these questions.

After being firmly endorsed by the Republican Party and assisted by its strong political reaches in Vermont, McCullough focused on what he thought would be the usual trot to the governor's chair in Montpelier. The Democratic Party was a distant afterthought.

The Democratic Party was a distant second in the Vermont gubernatorial races. A fringe party, the Anti-Prohibition Party, was a far-off third. Two decades of participating in gubernatorial races did not change this. As stumping season was about to commence, the Anti-Prohibition Party was in need of a champion to whom they could throw their support. The party found their candidate in none other than Percival Clement.

Clement was a native Vermonter born and raised in Rutland. This became a key factor, since General McCullough and Seward Webb, the Democratic Party candidate, were not native-born. This important detail was not lost on voters. After completing his education at Trinity College in Hartford, Connecticut, Clement settled back in Rutland to help operate the very successful family marble business. The family business was Clement's primary focus, along with his growing family (he had a total of nine children with his wife, Maria). But he also started to have some political aspirations. While still running the marble business, serving as the president of Clement National Bank and having many railroad investments, he served in the state house of representatives as the representative from Rutland from 1892 to 1893. It was with his political influence during his only term in the house that he was able to have the city of Rutland separated from the town of Rutland. After his house term, Clement was elected mayor of the city of Rutland in 1897. He again served only one term, then returned to his business affairs. His time away from the limelight was short-lived. He returned to Montpelier only two years later after being elected the state senator from Rutland County. After a single term, Clement wanted to aim for even a higher office—governor of Vermont.

During this period of his life, Clement invested an extensive amount of time, resources and personal fortune into the anti-prohibition movement. Just as the temperance movement had gained momentum earlier that century as a response to the excessive alcohol consumption that was dotting the landscape, the anti-prohibition movement was gaining extensive support across the state and rallying against the strict prohibition laws. While the support of the anti-prohibition movement was not strong enough to overthrow the political establishment, they were able to keep the topic of ending prohibition in the news and political conversations.

The temperance movement was gathering fervor from the religious movement known as the Second Great Awakening. At the same time, the anti-prohibition activists gathered support from the waning religious activism that emerged in the aftermath of the Civil War. A series of recessions and economic panics in the decades that followed also lent support. The financial instability in Vermont from these recessions and panics created a resistance to the political establishment. Coupled with the lingering effects of the "Long Depression" in the 1870s and early 1880s, Vermont's economy was hit rather hard. Some railroad companies and many banks collapsed. The loss of railroad development and construction stunted economic growth and left the state reliant on outside markets. This led to a general questioning of the state government's plan and laws.

As a very prominent member of the anti-prohibition movement with his political and business interests, Clement saw two primary issues with prohibition. First were the political ramifications that affected the state. Clement strongly felt that state prohibition, put into place by the Whig Party, was a gross overreach by state government into the daily lives of Vermonters. This was his first and most prominent platform. For this view, he was heavily lobbied by politicians, as well as citizens, to run as an alternative to the Republican Party's nominee, McCullough. Clement embraced the opportunity and declared his candidacy for governor, running as a Republican but on the ticket as the "High License Local Option" or member of the Anti-Prohibition Party. The second issue Clement saw was the fact that many towns and cities had relied on the revenue from taverns, saloons and alcohol producers. This formed a large portion of yearly municipal budgets. Once prohibition laws went into effect, there was a noticeable drop in available funds that created heavy deficits in many areas in the state. Some towns and villages, such as Salem, were unable to survive and went bankrupt. In the case of Salem, it was absorbed by the town of Derby, which created one of the largest towns (by land area) in the state.

The High License Local Option was the rallying point for the Anti-Prohibition Party—this was the belief that Vermont's state government should remove state-mandated prohibition and instead let local town and city governments decide whether they wanted to be wet or dry. If the town or city voted in favor of prohibition, the same law that was already in existence would continue to be enforced. If the city or town voted in favor of lifting prohibition, they would be allowed to issue a handful of liquor licenses within the town. Along with the restriction on how many licenses could be issued, the licenses carried a very high yearly fee. The revenue raised from the issuing of these licenses would then come back to the municipality for infrastructure development and improvement. This proposed plan had many supporters, and many voiced frustration about the dire need for infrastructure improvements, but not all said this possible new source of revenue for towns was a positive feature.

The approach that Clement took caught both the Republican and Democratic Parties by surprise. A serious disruption to the status quo had not occurred in politics in the fifty years that prohibition had been the law in Vermont. The heightened rhetoric in the gubernatorial race of 1902 was something out of character for the state. Reports in Burlington newspapers revealed a series of clubs that were organized to give away free beer and cigars in exchange for hearing speakers in support of McCullough. This was ironic, what with McCullough being the establishment candidate set on maintaining Prohibition. Yet, he offered beer as an enticement while Clement was the outlier promoting an end to the failed law. Two different reports about this are included below:

FREE BEER AND CIGARS
Political Club Rooms Practically Barrooms.
BARRELS OF POLITICAL BEER.
Club Houses Established All Over the City for Use of Members With Free Beer Galore.
[From the *Burlington Daily News.*]

There has been much comment of late regarding the club rooms which are in operation in this city supposedly in the interests of General McCullough. It would be especially interesting to Rev. G. W. Morrow, the Anti-Saloon evangelist, who pronounces the General as a "safe man," to look into the matter a little and see how many club rooms there are and how they are conducted. For the benefit of its readers and for Mr. Morrow the News will describe a few which are in active operation.

There is one at the residence, 33 North avenue. The tenant or owner is a German and this room is for Germans only. Two kegs of beer have been furnished weekly besides cigars. The owner receives $5 per week for the use of his rooms and there about 21 Germans gather to discuss the "Fatherland."

Just above this place on North avenue is another club room in the building said to be owned by a prominent citizen. French democrats and floaters attend here. No others need apply. At the beginning of the campaign "Lord Nelson" cigars were furnished at the patrons lived in clover. Now a cheaper brand is furnished which requires from ten to twenty matches to light, and a kick comes with every cigar. About a barrel of beer each week is drank by the members who further spend their time using the six card tables in the room.

There are two club rooms on La Fountain street, one at No. 45 at the corner of Rose street. The owner has always been known as a democrat. This club is known as ___ ____s McCullough Club room. It was 60 members, mostly French democrats and floaters. This club is allowed two kegs of beer each week.

Just across the street, in the Hamlin block (in the third story), is another. The room was formerly a garret, but it has been sheathed, painted and fitted with electric lights, chairs and tables.

The elite orate here on Saturday nights, after which beer galore is served.

At 19 North street lives a barber. Over the barber shop, one flight, with entrance from North street, is another of Mr. __ ___s clubs. Twenty dollars per month is paid. Beer and cigars are furnished to a better class of the French voters, but not much is said about it on the outside. The owner is an orderly citizen. Meetings are generally held on Saturday night. About 50 is the number of members.

Besides these are room in Parent's block, corner of Intervale and Archibald, and at Walnut and Archibald streets. In all of these club rooms there are not 50 republicans in attendance. We mean by this, men who vote the straight republican ticket without being paid for it. Most of them are floaters.

The German vote is not a purchasable vote as a rule, but they go with the party furnishing the most beer.

Not an Irishman can be found in the ward who is ready to admit he is a member of any of these clubs. When one of them presents himself at the door he is informed that "this is no Irish club," and the door is closed against him.

In all these club rooms, meetings will be held tonight to keep the voters from attending Mr. Clement's lecture. In other words, the lieutenants propose to use beer.

HERALD AND NEWS, *MAY 22, 1902.*

Cards, cigars and kind talk to keep the voters from hearing a simon pure local option speech. Very peculiar.

Another interesting illustration of the man's word to promote the "safe man's" interest is found in the meeting held a few nights ago. Mr. opened a room and invited the medical boys down to have a good time. About 75 went down and were regaled with two kegs of beer and cigars in plenty. Judging from the bottles lying around, other stimulants were there also. Speeches were made. Thus the boys were beered and enlightened. A News correspondent met one of the entertained reeling home thoroughly drunk.[31]

As the news reports show, Vermont was still awash in alcohol even though prohibition was on the books. Prohibition was also the focal point of the gubernatorial race. The reports generated passionate responses from residents and papers alike. One response was:

> *To the Editor of The Messenger:*
>
> *I would respectfully ask the sensible, temperance, fair-minded men of the state of Vermont, the fathers, (and mothers, too;) brothers, and friends of good citizenship if they will endorse the methods used for the purpose of nominating, McCullough, by voting for, or with men who seek to make proselytes by the following method, which is a scheme from hell, and an association with the devil :*
>
> *Within the past few days two well known attorneys in the city of Burlington called a goodly number of college boys from the U. V. M. to a rendezvous and delivered themselves in favor of McCullough. They smoked them with cigars, gorged them with beer, turned them out filled with vile stuff that makes a father mad, and mother sad.*
>
> *Up to this date I have not been radical on the governorship question, but from now on my labor and prayers shall be for the man who has the courage and purpose of F. D. Proctor.*
>
> *Temperance men of Vermont ! Line yourselves up ! "Choose this day whom you will serve." If you believe in the methods adopted by McCullough's agents in Burlington, be manly and vote for P. W. Clement. If you wish your school boys to be protected from the vampires of moral and social damnation, vote for Fletcher D. Proctor.*
> *Burlington, Vt., May 14, 1902.*
> *A Father.*[32]

Another response was:

> *The sober, order-loving people of this section will read with amazement and indignation the detailed story from the Burlington Daily News, published elsewhere, respecting the political club rooms that are conducted in that city in the interest of Gen. McCullough. The fame of these clubs has spread somewhat by current report, but this is the first description given in public print. Certain prominent McCullough men of Burlington have made haste to deny the responsibility of Gen. McCullough in the serving of beer at these club rooms, and say that Mr. Barber distinctly ordered that it should not be served. At the same time, it is established that beer has been served*

at some or all of these clubs. And in spite of their professions of innocence, the managers of the McCullough campaign, both state and local, cannot evade responsibility. They rented and fitted up these club rooms, at which smoking and card playing were established institutions. What else could have been anticipated than that drinking would accompany? The "beer vote" of Burlington and some other towns is not inconsiderable factor, and nobody supposes for a moment that the shrewd McCullough workers of that city would not take steps to land it in the most effective way, without, perhaps, giving the details to Gen. McCullough or those of his leaders who would object to such methods. There was no justification in the beginning for opening these joints, and those responsible for it cannot wash their hands of the consequences. It all comes of trying to use immense sums of money "legitimately," to accomplish political designs.[33]

Yet another:

Of the 19 Vermont newspapers that are supporting Mr. Proctor, only two advocate license. Of the 25 Vermont newspapers that are supporting Gen. McCullough, not less than nine advocate license. So much for the prattle about the "motley host" that follows Proctor's prohibition banner — and the silence about the composition of the McCullough following.
_____ *The McCullough people are welcome to what satisfaction they can get from their success in carrying six out of seven caucuses in Chittenden county, held last week. Chittenden county has been concededly favorable to Gen. McCullough, as a whole. In certain towns, where the McCullough sentiment was supposed to be so strong as to make the scheme reasonably safe, "snap" caucuses were called. The warnings for these caucuses were posted days before the official call for the state convention was issued—a thing unprecedented in Vermont political history. The caucuses were held within a very few days after notice was given, so as to afford little chance for opposition to organize. In every case, the down committee was controlled by the McCullough partisans, and the whole arrangement was doubtless planned out in advance by some central agency that pressed the button. The purpose, of course, was to create the impression throughout the state that everything was going for McCullough, and to cause a general stampede. In one town, this plan fell down. Essex elected a Proctor delegation, on a close vote. This was a stunner to the McCulloughites, and they raised the cry of fraud. It is alleged that 12 more votes were cast than were checked off the list. This might, and doubtless did, result from the fault of the clerk, but even*

casting out these votes Proctor had a safe lead in the caucus. The result will not be determined by seven towns, "bejiggered" in this way.[34]

The clubrooms of Burlington were handing out pints. Religious figures and leaders were hard at work stirring up support for the candidates. In a speech given in support of McCullough and in opposition to the High License Local Option in St. Johnsbury, a religious figure preached, "a man not altogether temperate said to me recently that old prohibitionary law has gone to the devil, hasn't it. Well, that were indeed deplorable but not half as deplorable as that the people of our state should go to the Devil also."[35] The speaker also pointed out "in his argument during the construction of the referendum, one man said he didn't want the money that was paid for licenses, to build roads with, for it was blood money anyways." Later in the speech, the speaker also noted that "another man said he didn't think most of the towns would care whether it was blood-money or not so long as they got it [money for infrastructure]." The speaker proceeded to be offended at such a suggestion, continuing with "can it be possible that there is a man or woman living within our borders who could take pleasure in riding over road, however so beautiful, when those roads are paved with the broken hearts of drunkard's wives, and the sighs and sobs of drunkard's children."[36] The fiery rhetoric continued. It was an attempt to shame anyone who thought the High License Local Option was a positive idea.

It is important to point out that St. Johnsbury was a prohibition stronghold. It was the home of the Fairbanks family, and it was Governor Erastus Fairbanks who signed prohibition into law. Most of the other cities in Vermont were in support of Clement, although the more rural areas were firmly in McCullough's grasp.

Clement announced that he would be running with Frank W. Agan of Ludlow. Agan was also a businessman and an inventor. He is best known for creating the Agan Vacuum, an early two-person vacuum cleaner. In his first political speech after announcing his candidacy for governor, Clement brought prohibition to the forefront. On the evening of April 12, 1902, at an overflowing town hall in Bethel, Vermont, he railed against the state legislature for their history of killing high license local option bills during a speech that lasted nearly two hours. The speech overexaggerated the status of prohibition and presented Vermont as being in a near-puritanical state.

The core of the issue, though, was that "the patronage of prohibition was part of the political machine and its position seemed impregnable."[37] Much of the revenue coming into towns and cities was from the enforcement of prohibition laws and the dubious penalties and fees issued.

The issue of fines and zealous prosecution over the decades also swayed Clement's views and fed into the speech he gave in Bethel. The start of his speech gave a brief general background of the history of Vermont's state prohibition and the aftereffects of the laws on towns and state politics. He then launched into what would be his core political message in the gubernatorial race. The core of his beliefs on Vermont's state prohibition was that:

> *the function of government is to control the individual as little as possible, allowing him the greatest freedom in all directions and only limiting his action where it interferes with the rights of others. We don't dispute that intemperance is a vice, but prohibition does not supply the remedy. We are not all endowed with the same moral perceptions. Thank god for that. And the question of what constitutes intemperance must be settled by each individual for himself. What would be intemperance in one person would be temperance in another. A man may go through his whole life eating moderately, smoking moderately, drinking intoxicating liquors moderately – in fact taking all the pleasures and business of life in moderation; he is a temperate man. On the other hand, a man may be a glutton, intemperate in nature and action, a crank in his ideas on all subject; but, if he is a teetotaler, the prohibitionist calls him a temperate man and denominates the other as being intemperate. The prohibitionist, while he is not always able to control himself even in the matter of drinking intoxicating liquors, seeks to control his neighbor, not by precept and example, not by argument and moral suasion-that process is too slow to suit his ideas of progress, and besides sometimes his neighbor tells him to mind his own business- but the prohibitionist seems to think it is his business to attend to that of his neighbor; so he gets the prohibitory law, and with that in his hand he looks over his neighbor's fence of a morning and says: "I've got you now where you can't drink whether you want to or not," and proceeds to tell him that it is a crime to be visited by penalties, only next in severity to those that are inflicted upon murderers, to sell a glass of cider, and that is a vice and sin, which will damn his soul forever to one drink.[38]*

Clement would continue to argue that the prohibition law that came into effect in 1853 was from a previous political landscape. That landscape had eroded away by 1902. Over the years, the original law had grown in reach and authority. Clement argued that it constrained the citizens of Vermont. A speech prepared by S.J. Beatty was originally to be read at the state gathering of Republicans but was suppressed until after the event. In it, he made a point supporting Clement's views that "unclean politics in Vermont are largely due to the demoralizing influences emanating from the prohibitory statute and it is the duty of every man who has the welfare of the state at heart to register his vote and use his influence to regenerate the state. A political and not religious revival is what Vermont needs."[39]

The second key point Clement made in his Bethel speech was about how Vermont's prohibition law dissuaded construction and investment in new hotels and different types of businesses. He carried this theme through his candidacy. He cited one incident in which an investor was prepared to invest $200,000 in 1901 to construct a magnificent large hotel on the waterfront in Burlington. After reviewing the laws, the investor informed Clement that he was looking elsewhere to construct the hotel. The reason was that he was "liable to 50 year's imprisonment for selling a bottle of wine" to his hotel guests. The investor ultimately constructed the new hotel in New York on Lake Champlain. Clement noted, "should you be in the upper part of the city of Burlington you can look across Lake Champlain and see the roof of his new hotel."[40] This example is one of many that showed the state losing revenue. It also raised a very important point—it was impossible to know how much the revenue lost during prohibition forced the shifting of Vermont's economy.

In six short weeks of political maneuvering and public speeches, Clement moved the issue of prohibition and the solution of the high license local option to the front and center. The political wrangling of Clement's platform meant that the issue of prohibition and outcomes associated with it would be the responsibility of the gubernatorial winner.

John McCullough, the mainstream Republican candidate, had taken up the High License Local Option cause within his own platform. Clement was more aggressive in the repeal of state prohibition. He was for letting the towns decide whether to be wet or dry and to handle the issuing of licenses. McCullough was more conservative. He supported the High License Local Option for political gain and not because of personal beliefs.

With significant attention to prohibition repeal at the forefront of the election, it was no surprise how close the results were. With the general election taking place on September 2, Clement had gained significant ground on McCullough. The results of the election were much different than in previous gubernatorial elections. The *New York Times* ran a story in the August 31, 1902 issue that had the headline "Campaign in Vermont; Bitter Contest in That State to End Tuesday. Fight Between McCullough and Clement Republican Factions—Legislature May Have To Name Governor." The article went on to state that the governor's race "was the hottest political campaign in the history of Vermont."[41]

With prohibition being the core issue of the election, the mainstream Republican nominee McCullough received 34,968 total votes. This was equal to 45.6 percent of the vote. Percival Clement got 31,864 votes, or 40.3 percent. Felix McGettrick, the Democratic Party's nominee, came in a distant third with 7,364 votes, or 10.5 percent of the vote. The results did not present a clear winner in the election. State election rules required a 50 percent plus one vote majority to win. These results meant that the legislature would decide the election when it convened the following month.

The legislature came to a decision on the first ballot taken. McCullough was voted in as governor with 135 votes, which equated to 61.2 percent of the vote. Clement received 59 votes, or only 22 percent, and McGettrick got 45 votes, or 16.8 percent. It came as no surprise that McCullough was voted in as governor as the Republican Party–backed candidate. Clement was running as a third-party candidate. Many newspapers leading up to the election had speculated that Clement's only real chance at gaining the governorship was winning the general election outright. The difference between the general election and the legislative votes was striking. The general vote results showed that while McCullough pulled most of the Republican vote, Clement pulled votes from across party lines. Clement pulled support from those opposed to prohibition, along with some Republican and some Democratic votes. The core of McGettrick's support was some of his own Democratic vote and much of the Prohibition Party. The Prohibition Party had an outspoken dislike of both Clement and McCullough. As a Republican, McCullough was forced to take a mild stance on the repeal of prohibition out of fear of losing Republican votes to Clements.

The legislative vote was far more straightforward, with the Republican Party throwing all its support behind McCullough. The divide in the legislative vote was more between Clement and McGettrick. Clement had received more than four times the votes McGettrick received in the general election, yet he got only 4 percent more than McGettrick in the legislature. The vote was most likely reflective of how Clement's rhetoric in his campaign irritated many in the Republican Party. It was the first election since 1854 that the Republican nominee did not win with a clear majority.

Similar results occurred in the lieutenant governor's race. Republican candidate Zed Stanton received 47.2 percent of the general vote to Frank Agan's 38.1 percent. Elisha May, the Democratic candidate, received 11.4 percent. When the legislature voted, Stanton received a clear majority, winning the seat.

Even though Clement lost, he had managed to gain enough support and attention that he left Governor McCullough in a complex web of issues. In McCullough's inaugural speech he gave to a joint session of the legislature on October 3, 1902, he wasted minimal time tackling the issue of prohibition. After a formal acceptance of being named governor and praising the state's tried-and-true political system, the first issue he tackled was prohibition. He announced that the people had spoken, and that the future for the state was to move away from complete prohibition to a restrictive High License Local Option. In his speech, McCullough stated that:

The verdict of the freemen of the State on September 2d last was in favor of the General Assembly framing a local option and high license law and submitting the same to the people for their adoption or rejection. This duty will require the very best efforts and the most intelligent consideration of the members of the Assembly. For fifty years prohibition has been the policy of the State. The mandate comes up now from, the people to their legislators commanding them to formulate and to submit to them for their decision some other system.

This, the Anglo-Saxon, the American method. It is the rule, of the majority. And primarily, on this subject, it must be borne in mind that all sumptuary legislation must be supported by public sentiment to be effectual. In framing a statute the General Assembly will have the benefit of the legislation on the subject of eight or ten of the other States. Experience is the very best guide. In every State the difficulties arising from the United States internal revenue laws, from the freedom or interstate commerce guaranteed by the National Constitution, and from the medicinal and industrial demand for alcohol will always embarrass the enforcement of any law.

Different States and different parts of the same State may require different treatment, Unlike most of the States, Vermont has few manufacturing centers or large municipalities; the great majority of her towns are rural and agricultural. Massachusetts is more nearly similar to Vermont than any other State and from her legislation probably more valuable suggestions will be derived than from any other source. But in the legislation of no one

State, only, should we look for the best and wisest provision and those most suitable to the circumstances of our people.

In any local option or license system, it is worthy of consideration whether the vote on License or No License should be taken in any town or municipality oftener than once in three or five years; whether it should not be taken at elections specially called for that purpose and not at any regular election. State or local; whether if License be voted, it would not be wise to require the petition of a majority of the property holders in any block or square of a municipality before issuing a license whether a majority of the legal voters of any town or sub-division of a city should not be allowed to remonstrate against licensing or continuing the license of a specified person whether any license should be granted for more than a year; whether the number of licenses where authorized should not be limited to one for every 1,000 inhabitants, and prohibited within a limited distance of any church, school house, theater, opera house, public building, park or other public place; whether all licensees should not be required to give am bonds, and every applicant for a license furnish evidence of citizenship and good character. As to the licensing body or authorities, it has been well said that judicial purity and reputation for purity are far more important than discreet licensing.

It is of the utmost importance that courts and judges should be kept as far removed from politics as possible. This matter of licensing therefore should be entrusted to some other department or to boards specially raised up for that purpose, and which boards should have stability and independence. There should he several grades of license fees depending on the size and population of the towns or cities; and the traffic should be made to raise large revenues for both the State and the towns or municipalities.

Every licensee should be restricted from selling to minors or intoxicated persons, or on Sundays, election day or any legal holidays, nor should he be allowed to furnish musical entertainment of any kind or billiards or cards or any game whatever; and the place should be wide open to inspection from the street or highway and the hours should be strictly limited, and the shorter the better, provided public sentiment supports these restrictions.

If druggists' licenses are to be granted at all, they should be entrusted only to registered pharmacists who should be authorized to sell only in small quantities and only on the written prescription of a physician not interested in the store. These suggestions, gathered from many sources, may be of some value in formulating a proper statute to be submitted to the people for their adoption or rejection.[42]

This speech laid out a rigid framework for how the proposed law would replace the laws of prohibition. Through debate in the house and senate, a high license local option bill was crafted, and on December 11, 1902, Act 90 was passed. Entitled "An Act To Regulate the Traffic in Intoxicating Liquor," it gave towns the right to vote at their annual town meeting whether to be wet or dry. The act also gave towns control over what kind of license they would issue if they voted to be wet.

Also important is how the act was presented to the public. The legislature decided that Act 90 would have a public referendum scheduled for February 3, 1903. The referendum would allow voters to pick the date that Act 90 would go into effect. A yes vote in February would allow Act 90 to go into effect on the second Tuesday in March of that year. A no vote would mean that the act would go into effect the first Monday in December 1906. The formulation of the referendum was to allow voters to choose the day the implementation of law would begin, not a choice about the law itself. The option of the no vote to give three years for the act to be enforced would give the legislature time to find a viable alternative and repeal the act. The February referendum vote resulted in a yes vote, 29,711 to 28,982.

Between the passing of Act 90 and the February referendum, Vermont looked like it was in the midst of another political campaign. Posters and broadsides both for and against the act were everywhere. Many groups rallied their members to vote one way or another. One slogan that popped in defense of prohibition was:

> *Men of Vermont, be loyal and true! Vote against the saloon. Remember the wife; remember the mother; remember the sweetheart; remember the daughter. They are asking you now to stand by them–to leave the rum and stand by them. Not one is asking, but all are asking–every woman in our fair State! Does she not care? She does care, for none knows better than she what wrecks the home, what separates the hearts, what brings poverty, suffering, crime. Green Mountain Boys, do your duty; vote against the saloon.*[43]

Other slogans were in support of repeal of the high license. Newspapers had headlines either in support of or against the act. In the short time until the vote, both sides moved quickly to spread their messages.

A rallying cry for those in support of the High License Local Option was the issue of revenue. When looking at fifty years of prohibition in Vermont, the loss of revenue and the economic damage is grossly understated.

The revenue that left Vermont also left the state's economy. This caused far more reliance on agricultural and business pursuits, and these were unable to make up the gap in losses. Ironically, at the time that state prohibition took effect in 1853, one of the leading agricultural crops was hops. The production of the hops industry in Vermont peaked in 1860. At seven years into state prohibition, the timing in peak production was remarkable. Only one brewery was producing beer in the state. The total amount of hops produced in 1860 in Vermont ranked second in the nation. This created a much-needed lucrative crop for the state.

But three factors would result in the demise of this lucrative cash crop. The onset of the Civil War eroded national brewing capacity and the need for hops. There was a series of difficult growing years in the mid-1860s. And there were the added costs of sending the entire Vermont crop to other markets. Within a decade, the industry declined. In fewer than twenty years from the onset of prohibition, Vermont's hops industry was no longer commercially viable. This led to an agricultural and, more importantly, financial void for famers who were reliant on the cash crop. Prohibition's direct damage on Vermont's hops industry was remarkable. It would take over a century before the crop returned to be barely a commercial viability on a significantly smaller scale.

It was not only Vermont's hops crop that was hit hard. Alcohol production was heavily taxed by both federal and local governments. This resulted in a revenue stream. While taxation of alcohol production caused a stir with brewers and distillers, they never forcibly challenged it. The loss of production in Vermont meant that revenue sources for the state dried up overnight.

Another lost revenue source was tourism. With all surrounding states and Canada being wet, the 1853 law put Vermont at a gross disadvantage for generations. While portrayed as picturesque and an outdoorsman's paradise, the fact that it was dry sent tourists elsewhere. In response, some pushed the idea of temperance houses and hotels as family-friendly destinations—but those did not draw visitors. Not all businessmen were willing to stand by the wayside of Vermont's degraded hospitality industry. In some cases, the price one had to pay for getting caught serving alcohol did not matter.

One of the largest and grandest hotels in Burlington was the Van Ness. It was at the corner of Main and St. Paul Streets during the late nineteenth and early twentieth centuries. Urban A. Woodbury purchased the hotel in 1881 and was a leading citizen of Burlington. After purchasing the hotel, Woodbury was elected mayor of Burlington and served from 1885 to 1886.

He served as lieutenant governor in 1888 under William P. Dillingham and was ultimately governor, an office he held from 1894 to 1896. Throughout his political endeavors, Woodbury oversaw his hotel. He defiantly and openly served alcohol at the hotel, even during Vermont's prohibition. With his political connections, not to mention a few American presidents as hotel guests, Woodbury had few legal problems. Many other proprietors in the state were not as daring or open about serving alcohol.

Many towns and cities in Vermont had concerns about the high license local option. If a saloon should open in their town or a neighboring one, where would the money go? Joseph Nelson Harris of Ludlow printed a pamphlet in full support of McCullough for governor over Clement. This pamphlet also looked at the revenue effects of the high license local option at a local level. In his pamphlet, he noted that:

> *Ludlow has a population of about 2,000, with about 600 voters, a little less than one in three. It was shown in our Republican town caucus, that about one-third of the entire population gave Clement a majority, and were in favor of local option. With such a law, to work on the basis of similar laws in other states, Ludlow would have two saloons, and the price of the licenses would undoubtedly be fixed by a county commissioner and would perhaps range from $250 to $1000 each, according to whether it was a first or second grade license, and out of this sum the state would be entitled to a commission ranging from fifteen to twenty-five per cent. It does not seem as though Ludlow could afford to have saloons on the basis of the income derived from two licenses. What profits could be derived from the liquor traffic in this town, would illustrate those of every town in the state.*
>
> *Many of the voters have the idea that this optional whether we have a saloon in our town or not. That is very true, but if we vote no license and Cavendish votes for a license, the liquor customers of this town would give the benefit of their money to Cavendish or, if the case were reversed, the people would pay their money to our saloon-keepers.*[44]

Of the many statements on the high license local option, few grasped what Harris had published. Harris looked at the effect on local economics. He did not dive into the religious or social implications. Many towns launched campaigns against saloons in their community, while other towns campaigned for a saloon.

The resulting vote by towns in the referendum on Act 90 resulted in a cultural split in Vermont. A resounding majority of the rural towns voted in favor of the delayed 1906 date, signifying a desire to delay the implementation of Act 90. Towns with larger populations voted in favor, and with a higher number of votes, Act 90 went into effect in 1903. This vote also reflected how the 1902 law benefited larger towns and cities. The law started with one license per every one thousand residents. This meant that many of the smallest towns in the state were not permitted to get and distribute licenses. The most populated areas were able to have permits for multiple establishments and greater tourism potential. Not all population centers made the same choice. St. Johnsbury, a stronghold of the Fairbanks family, opted to institute the dry option for years after the high license local option gave localities a choice. For Burlington, the option to be wet meant that the city could compete in the hospitality industry.

The results of the election got noticeable attention in the press. McCullough had won the seat for governor, and he is credited with the formal removal of prohibition. Percival Clement was the recipient of praise for dislodging the law that had been in effect for a half-century.

PROHIBITION SUPERSEDED IN VERMONT.
AFTER clinging to prohibition of the liquor traffic for half a century— for a longer continuous period than any other State—Vermont has repealed the law and adopted in its place local option and high license. Fifty years ago the prohibitory law was adopted by popular vote by the slender majority of 1,171, and last week it was voted down by a majority of about 1,600. The new system will take effect in March. "Vermont by its vote raises itself out of a mire of hypocrisy," says the Brooklyn Eagle, but the Providence Journal remarks that "in a State which has given a Presidential candidate more than 40,000 plurality, the majority of 1,600 for high License looks paltry enough, and must be taken as an indication, that the issue is still open." The Burlington (Vt) Free Press, which opposed license, declares that "the one gratifying feature for the opponents of the new law is that the margin in its favor in the popular vote is so small that its advocates will recognize their system to be on its good behavior, and the abuse of license will certainly cause a revulsion in popular feeling sufficient to overthrow it."

Some papers remark that the real test of the high-license sentiment in Vermont is yet to come. The real sentiment will be revealed next month, when each town and city shall settle for itself whether it will have license or local prohibition. Says the Rutland (Vt.) Herald.

"The dark night of prohibition is over. Vermont, this morning, is a part of the great world of intelligence, commerce, and wealth. We are no longer a peculiar people. We would be foolish to anticipate an easy road at first. The transition from one system to another is always attended by friction. To change the figure of speech, you can not make an omelet without breaking eggs, and you can not right a wrong without a disturbance. But we believe that there is stamina enough in the Vermont character to meet the responsibilities of new duties to which it is called with wisdom and self-restraint."

Much credit is given to Mr. P. V. Clement, of Rutland, for the result. He started the agitation against prohibition last fall, had a pledge put in the Republican platform to submit a new measure of this kind to a vote of the people, and then ran as an Independent Republican against Gen. J. G. McCullough, Regular Republican, who was elected by a small majority. Mr. Clement stood for high license and General McCullough for the prohibitory law. The Springfield Republican comments:

"Prohibition is still available wherever wanted, and probably two-thirds or more of the towns of the State will hold to it at the coming town-meetings. The cities and large towns, where liquor has been sold freely and often openly, in defiance of law, will generally choose a license policy, which will simply mean a regulated, revenue-producing traffic in liquor, in place of an illicit, lawless, and demoralizing sale conducted on a scale that the license policy will find it difficult to exceed. A system of high, restricted license like that of Massachusetts has been provided, and that it will prove as satisfactory in Vermont as it has in this State, we have little doubt. Anyhow, those places which do not want it need not have it, and in the application of this principle of local option is to be found the surest method of dealing with the liquor question."

The law provides seven classes for licenses, to be granted at fees ranging from $1,200 for a saloon to $10 for a druggist, who 'can sell for medical purposes only. Each town voting for license may have one open bar for each 1,000 population, exclusive of licenses granted to summer-hotels and to drug-stores. 'With 'Vermont out of the prohibition column, there remain but four States in which prohibition prevails, namely, Maine, New Hampshire, Kansas, and North Dakota, and in the second of these -only the sale at retail (not the manufacture) is prohibited.[45]

The original high license local option referendum and Act 90—as proposed by McCullough and passed by the general assembly—did not last long. Act 90 passed in 1902 and was removed in 1904. It was narrow in scope and limited the power to issue penalties. Act 115 of 1904 became more restrictive in prohibition laws. It substituted new, broader-reaching laws while still keeping the high license local option in place and the conversation about prohibition ongoing.

One of the changes that occurred between 1902 and 1904 was the length of sentencing for those convicted under the law. From the 1902 statute, the sentencing was limited to a maximum of three years of imprisonment. There were guidelines in place for judges that stretched back decades. The 1904 revision removed the limit on punishment with no replacement. Now, judges had the ability to impose fines and jail time at their discretion. The new statute took effect on March 1, 1905. The move was an attempt to strengthen the law in the public's eye. There was also an "alternate sentence" option. A judge could issue a fine for a violation of the law. If the guilty party could not pay the fine within thirty days, the monetary fine became the duration of a prison sentence. Many times, the individual was serving in prison over those thirty days. After this drastic change, some cases were appealed solely based on cruel and unusual punishment.

This change was showcased in *State v. G. Ceruti* in Caledonia County court in December 1909. Ceruti was caught in possession of liquor at his home. In court, he was fined $300 for his initial guilty plea, which he attempted to change to a plea of not guilty. Ceruti thought he would get a small fine. That fine would have been less than the cost of a lawyer and proceedings. He also thought it would remove his prison sentence, anticipated to be between three and four months. If he did not pay the $300 fine, the alternative sentence could be up to 1,200 days—or more—in prison.[46]

The high license local option lessened the former laws of prohibition. However, there was concern among lawmakers in Montpelier. Some believed that the sale of liquor would deteriorate the social fabric created from prohibition. The overhaul of the high license local option put more deterrents in place for potential lawbreakers. In doing so, the legislature succeeded in ending state prohibition in name only. By widening the legal ramifications and placing heavy restrictions on obtaining a saloon license, prohibition was still effectively in place for much of rural Vermont, where many communities never had a saloon reopen. The population wasn't big enough in rural areas to meet the 1,000-person criteria for one saloon. Other towns and some cities had strong local temperance movements that were so entrenched that a vote never passed in favor of issuing a saloon license.

One change from McCullough's Act 90 to Act 115 was more governance on the county level. Each county had a county liquor commissioner who issued saloon licenses. The cost of licenses increased. It did not matter if you had a class one license as a full-fledged saloon with on-site consumption or a license to merely sell only lesser beer and hard cider. The increased costs forced some saloons to close. They were not commercially viable if a neighboring town also had a saloon. In some cases, towns wanted no part in the evils of having a saloon.

The comparison of statistics of arrests for intoxication before Act 90 and after show another story. In many counties, arrests climbed during the infancy of the high license local option. Some counties, such as Addison, had very few arrests over the four years before the local option passed; Addison County had 14 arrests in that time, but in the four years after the option took effect, there were 53 arrests in the county. Chittenden and Rutland Counties had the biggest swing in arrests. Chittenden County had 545 arrests, while Rutland County had 396 arrests over the last four years of prohibition. Once the high license local option passed, Chittenden rose to 1,917 arrests, and Rutland increased to 1,147. In the case of Chittenden County, the swing was an approximately 400 percent increase in arrests. On the flip side, Washington County had the most arrests during the last four years of prohibition. After 485 arrests over those years, the county saw 544 arrests during the first four years of the local option.[47] These statistics fueled many temperance groups, although they did little to show larger problems with the local option.

Over the next decade, the debate between high license local option or prohibition would play out in the court of public opinion in Vermont. Some towns flipped back and forth. Others steadfastly remained in one camp or the other. The political landscape had seen a radical change after Percival Clement's first attempt at winning the governorship, and in 1918, he finally won the seat that had eluded him in 1902 and 1906. This win came just before the start of federal prohibition under the Volstead Act.

PART II
A CENTURY INTERLUDE

INTERLUDE

After the closure of the Burlington Brewery around 1885, and with prohibition lasting long enough to make a generational imprint, there was a sizable uphill battle to open a new brewery in the Queen City. Without any local production, Burlington drinkers turned to the beers available in New York and Quebec.

Perhaps surprisingly, in 1935, the first operation to produce alcohol in Vermont after the end of federal Prohibition was Green Mountain Distillers located on College Street. Utilizing maple syrup and sugar, they made "Colonial Liquor," a spirit akin to a sweet rum. This operation was short-lived, and part of the reason for its demise was the high cost of maple syrup. This practice has emerged once more in the state, with three distilleries producing vodkas from maple syrup: Vermont Spirits' **Vermont Gold** (Quechee), Caledonia Spirits' **Barr Hill Vodka** (Hardwick) and Elm Brook Farm's **Literary Dog** (East Fairfield). A white rum called **Pirate Dan's Vermont Rum** is now also produced by St. Johnsbury Distillery, formerly known as Dunc's Mill.[48]

Another aspect that altered the course of beer history in the entire country was the rationing of grain during World War II. While the soldiers had beer, its production was hampered by the grain rationing. Brewers, in turn, modified recipes for old favorites or drafted entirely new products. The addition of corn and rice to lagers resulted in paler, lighter-bodied beers, which the soldiers grew accustomed to; they brought home these preferences after the war. A similar situation occurred in England, resulting in a shift in beer ingredients and preferences. Some of the recipes and beers created during World War II are still widely consumed today.

During the century between the closing of Vermont's last historic brewery and the resurgence of the brewing industry in 1986 with the opening of Catamount Brewing Company in White River Junction, there was an onset of laws regarding alcohol production, the serving of alcohol and its consumption. In the 1960s, a law was enacted that barred patrons from having more than one alcoholic beverage in front of them at a bar. While this may seem innocent enough (and perhaps was intended to promote moderation), the result was that restaurants and bars were unable to serve patrons flights, or multiple samples, of craft beers. This law was changed in 2014 to allow for flights. Another law put a cap on the alcohol content of beer. In 2008, Vermont brewers ultimately persuaded the state legislature to raise the cap to 16 percent ABV, allowing brewers to explore a wider range of styles. Had it not been for these brewers, your favorite Vermont-brewed double IPA would be qualified as liquor and would only be available in state liquor outlets.

As a state that didn't have the type of real brewing heritage found in larger cities but did have a long history of religious decrying of alcohol, the current resurgence of brewing in Vermont is nothing short of remarkable. For Burlington, the craft beer renaissance has earned the Queen City an international reputation for outstanding beer and as a culinary paradise. Now, more than one hundred years after the city's first brewery closed, we present the story of brewing in today's Burlington.

PART III
BREWERIES AND BREWS

VERMONT PUB & BREWERY

144 COLLEGE STREET, BURLINGTON

Vermont's Original Craft Brewpub

In the 1970s, the craft beer movement was just an idea. Fritz Maytag purchased Anchor Brewing Company in San Francisco in 1965 and became a forerunner in changing the public perception of what beer could be.[49] Homebrewing was still illegal—a hangover from federal prohibition. But that didn't stop hundreds of Americans from brewing beer at home. One of these lawbreaking young lads was from Springfield, Massachusetts. Gregory John Noonan and his wife, Nancy, would go on to open the first brewery to operate in Burlington since the 1880s.

A SHOEBOX

By the time President Jimmy Carter legalized homebrewing in 1978, Greg Noonan was well on his way to becoming an expert in the field. Owing to the lack of resources available to small-batch brewers, Noonan started to approach his homebrewing scientifically. In the early 1980s, while living in Williamstown, Massachusetts, he started to compile research on best practices for brewing small-batch lagers, and it was turning into a book. In a letter to his friend Steve Polewacyk, he joked that he was having to relearn high school chemistry for his chapter on water.[50] His research into water

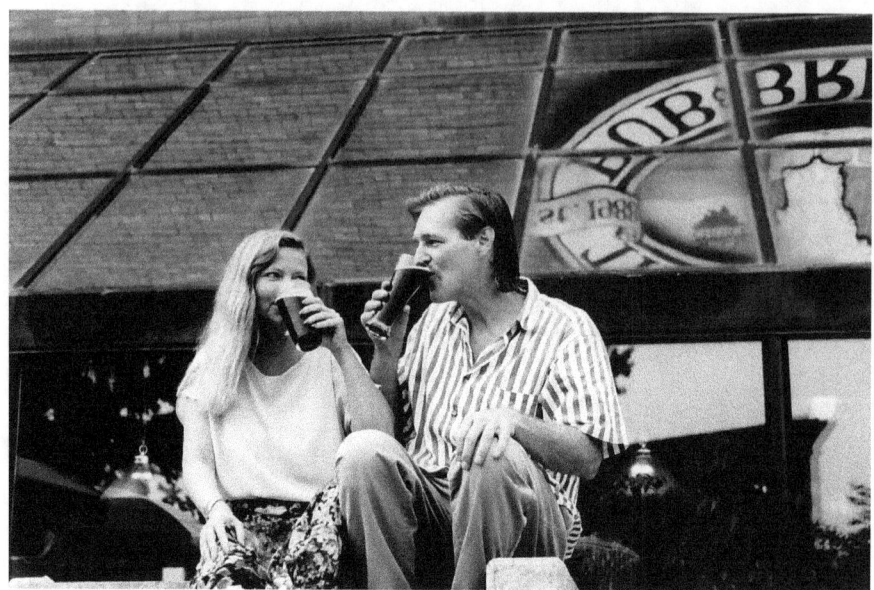

Above: Nancy and Greg Noonan
enjoying a pint on the patio at
Vermont Pub & Brewery. *Courtesy
Vermont Pub & Brewery.*

Right: The original Vermont Pub
& Brewery crow painted by Greg
Noonan. *Courtesy Vermont Pub
& Brewery.*

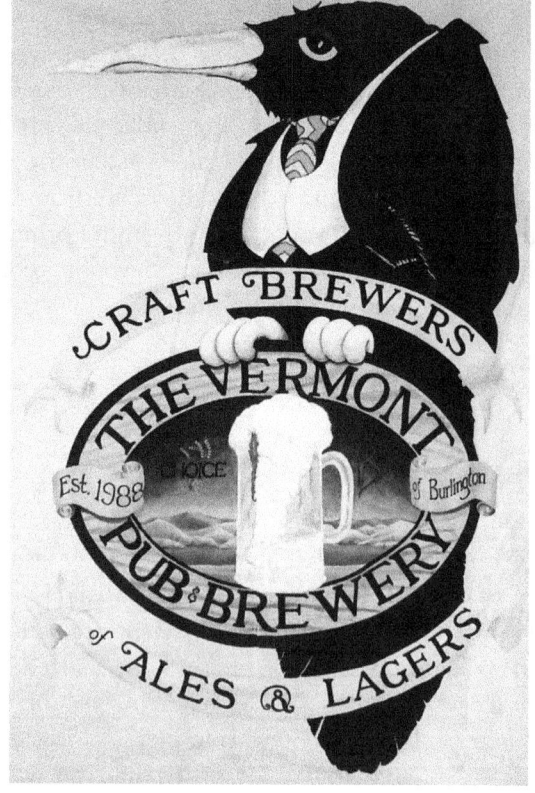

chemistry paid off, and writer Michael Tonsmeire called the water chapter "one of the most comprehensive" writings on the subject, as "it covers both the scientific (bonding, anions etc.) as well as the practical side (how to read a water analysis report from your water department or a lab)."[51]

In 1984, Noonan handed a copy of his manuscript to Charlie Papazian, the founder of the American Homebrewers Association. The manuscript was in a shoebox—an inauspicious and humble beginning for a book that would influence almost every amateur and professional brewer in the country. According to John Kimmich, "You're not going to find a successful brewer in the country that doesn't have a dog-eared copy of this book."[52]

Brewing Lager Beer: The Most Comprehensive Book for Home—And Microbrewers had its first printing in 1986. While writing this book, Noonan began to form a plan with his wife, Nancy, to open a brewpub on the East Coast. Brewpubs were scarce at the time—so much so that when he pursued loans, bankers asked, "What is a *brewpub?*"[53] After scouting locations in Massachusetts, New Hampshire and New York, the Noonans found friends in Burlington who convinced them that the Queen City was the perfect spot... well, almost perfect; they would spend three years lobbying to have laws and regulations changed to allow for the sale of alcohol for consumption on the premises where it was brewed.

GREG FOUGHT THE LAW

The Noonans made contact with a college friend, Peter Clavelle, who was, at the time, running Burlington's Community and Economic Development Office. Clavelle helped confirm their decision that Burlington was the right location, but there were legal hurdles. Noonan enlisted the help of homebrewing State Representative Bill Mares. Mares and Noonan spent years working on new legislation and had to convince lawmakers that these changes would have a positive economic impact on downtown Burlington.

> *"We had to get the law changed, to allow beer to be sold on the premises of the brewery. We spent 3 years on that, and it was an education. Vermont Legislature is government the way it should be done...They look at all the possibilities, all the implications, and make darn sure they know the ramifications of what they're voting on."—Greg Noonan*[54]

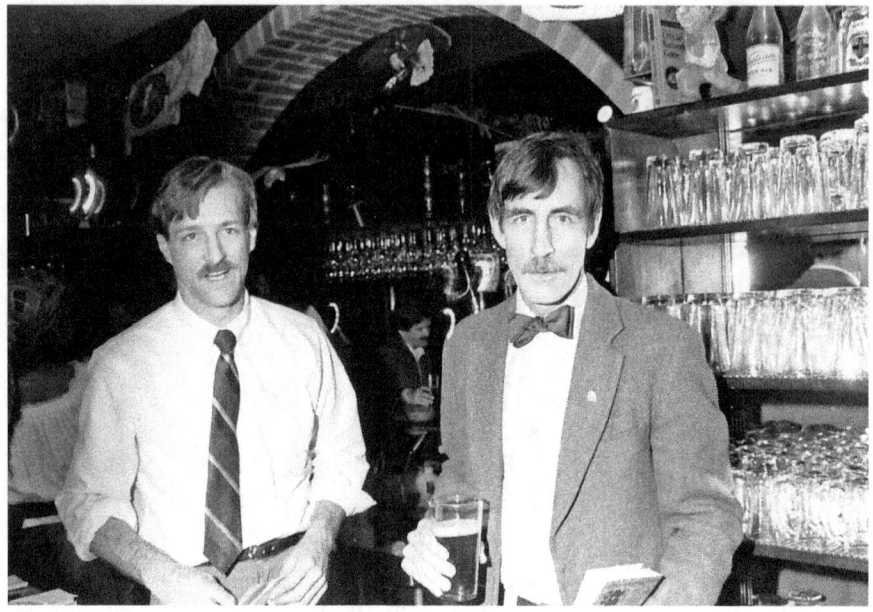

Greg Noonan (*left*) and Bill Mares (*right*) worked diligently to change many Vermont laws that prohibited brewpubs. Mares is holding a copy of his book, *Making Beer*. *Courtesy Vermont Pub & Brewery.*

On May 18, 1988, Vermont governor Madeleine Kunin signed into law several bills that would permit the existence of brewpubs, and on November 11, 1988, Vermont Pub & Brewery (VPB) opened its taps to the public for the first time.[55]

A SHOESTRING

With the legal framework sorted out, the Noonans had some work to do. They may have only had a shoestring budget of $179,000,[56] but Greg was a skilled draftsman and amateur inventor. Channeling Dr. Frankenstein, he built a fourteen-barrel brewery out of a maple syrup tank, a commercial ice cream machine and a pig feeder.

It wouldn't be an exaggeration to say that the Noonans were scrambling with the buildout of the brewery and the pub. Word spread through his circle of friends, and some of them flocked to Burlington to pitch in. Steve Polewacyk had become friends with Noonan via a mutual friend, and he

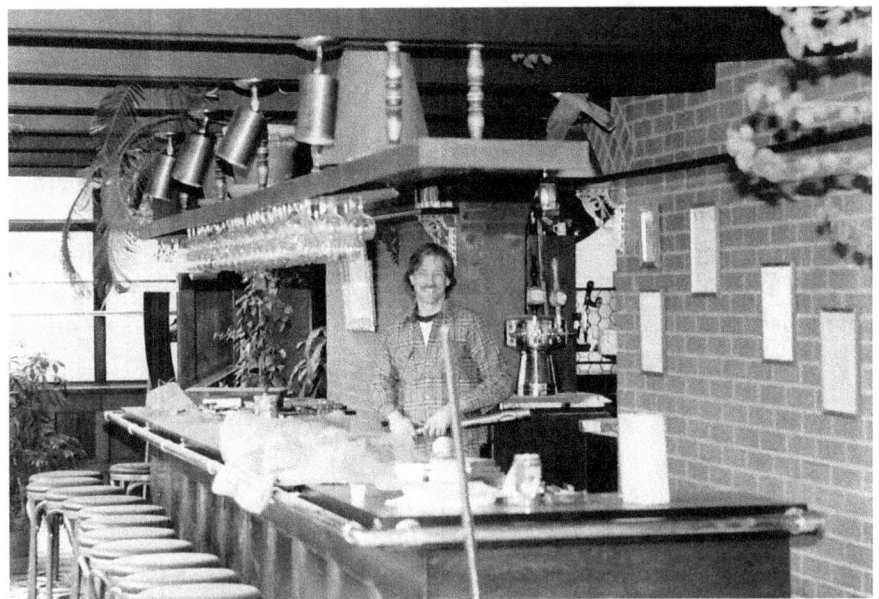

Greg Noonan smiles for a photo while putting on the final touches right before opening the doors at Vermont Pub & Brewery. *Courtesy Vermont Pub & Brewery.*

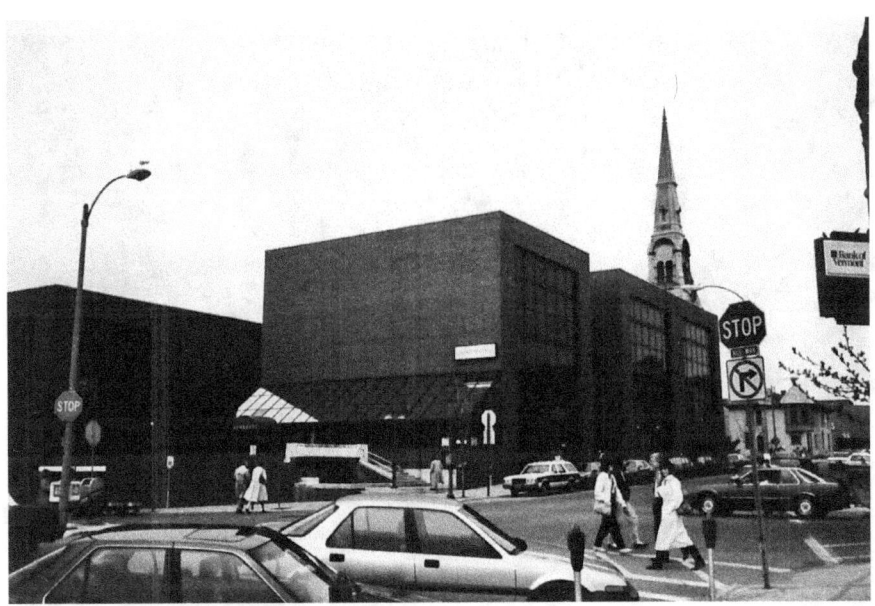

The building at the corner of St. Paul and College Streets was home to many restaurants, including Central Station. This site would become the home of Vermont's first brewpub. *Courtesy Vermont Pub & Brewery.*

decided to take a break from his rat-race job as a "computer guy" in New York City to help out his pal.

In the first two weeks that VPB was open, the Noonans and their team worked countless hours, only taking short breaks to sleep in the booths. Plumbing, tiling, steel fabricating, brewing, cooking, serving beers—after realizing how deep in the weeds the Noonans were, Polewacyk decided to permanently move to Vermont and become more involved in the business.[57]

After a few years of hard work and very little sleep, VPB had become established as a fixture in the Burlington community and would go on to influence the entire Vermont brewing scene.

AHEAD OF THEIR TIME

"In 1988, there were 22 recognized styles," recalled Polewacyk. Although the sky is the limit today, in the late 1980s, beer drinkers had only been exposed to a narrow selection of styles. Early patrons would ask what kind of beer they brewed here: "Budweiser? Miller? Coors?"[58]

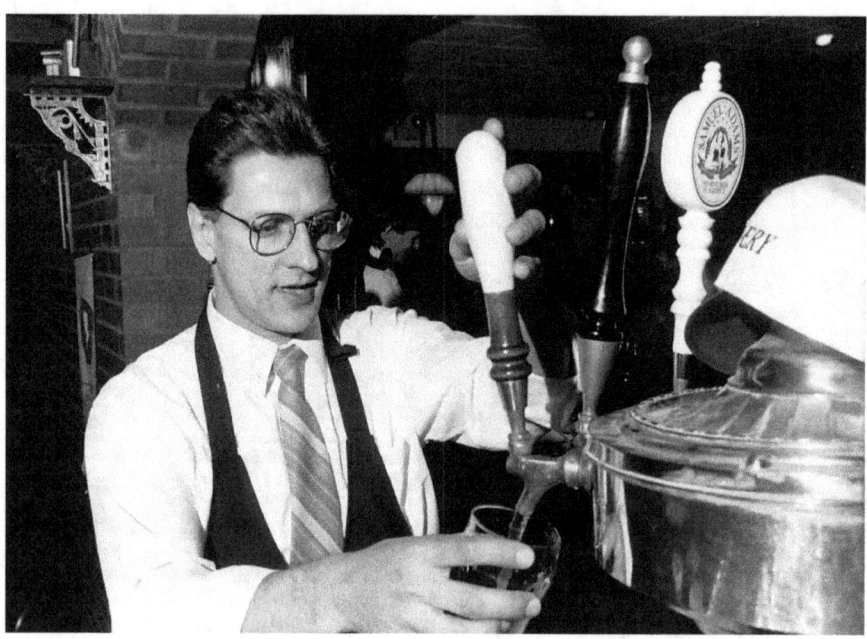

Steve Polewacyk pulling a beer at Vermont Pub & Brewery. Greg Noonan can be seen smiling over Polewacyk's shoulder. *Courtesy Vermont Pub & Brewery.*

This lack of exposure presented an opportunity to brew something new but also a challenge of educating patrons about new styles. The VPB team quickly designed a beer menu that included descriptions and historical facts about each beer. This was especially helpful when VPB would brew something that was way ahead of its time.

However, before there can be many, there must be a first. **Pilot Ale** was a low-alcohol beer—"to demonstrate our commitment to the responsible enjoyment of malt beverages"[59]—brewed with Pale, CaraPils, Munich and Cara40 malts and Willamette hops.

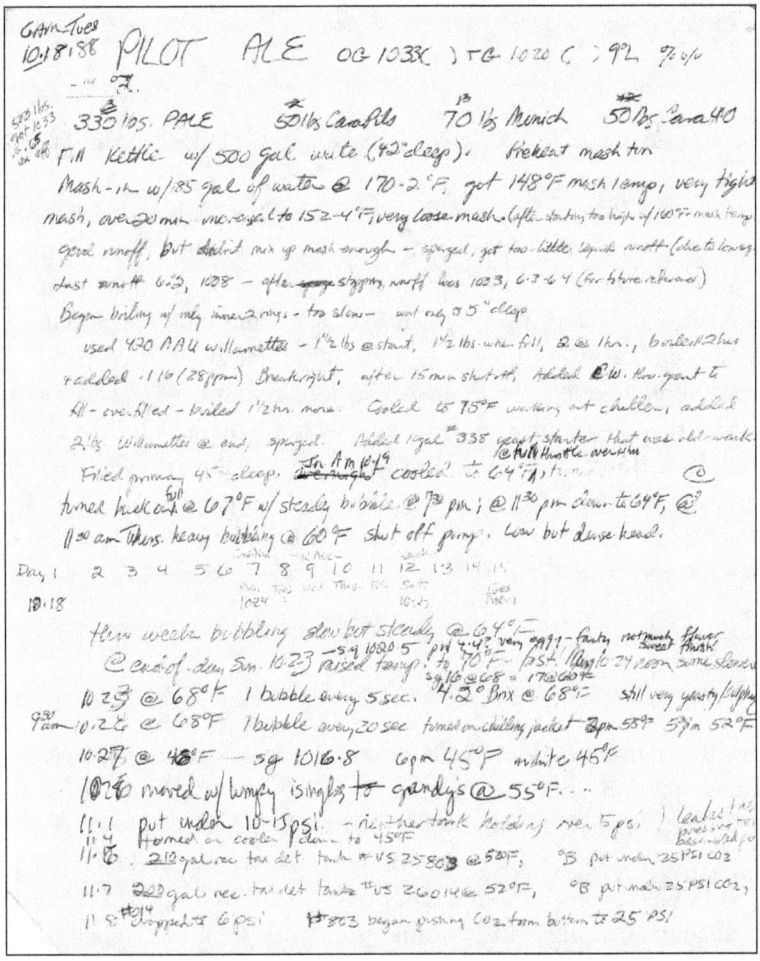

Greg Noonan kept careful brew logs at Vermont Pub & Brewery. The appropriately named Pilot Ale was the first beer he brewed once the brewery was completed. *Courtesy Vermont Pub & Brewery.*

In addition to presenting **Pilot Ale** on opening weekend, VPB also offered **Pesky Sarpent**, an unfiltered lager, and **Burly Irish Ale**, a sessionable Irish-style red ale. A guest tap featured **Catamount Brewing Company Porter**.[60]

Early on, Noonan decided to brew a beer with spruce sap, a tradition that had almost been forgotten.[61] But brewing with sap turned out to be a sticky endeavor. Noonan tried multiple emulsifying techniques but eventually decided to use spruce tips instead. Other Vermont breweries would eventually pick up the spruce-tip concept. In 2013, VPB collaborated with Lawson's Finest Liquids (based in Warren, Vermont) on a **Vermont Spruce Tip IPA**.

SEVEN BARREL BREWERY & AMHERST BREWING

In 1994, with VPB starting to gain steam, the Noonans founded a second brewpub just across the state border in West Lebanon, New Hampshire, calling it Seven Barrel Brewery. John Kimmich, who would go on to found the Alchemist Pub & Brewery, got his start waiting tables at Seven Barrel. When Noonan's assistant brewer Glenn Walter left in 1995 to open his own brewery, Noonan offered Kimmich the job. Kimmich brewed for a few years at VPB and met his wife, Jen, while working there. After the Noonans' divorce, Nancy ran Seven Barrel until she sold it in 2011.[62] Never resting on his laurels, Greg cofounded Amherst Brewing Company in Massachusetts in 1997, expanding the presence of brewpubs throughout New England.

GREG NOONAN

Noonan was much loved and celebrated both locally and nationally for his innovation, passion and—perhaps most of all—his willingness to talk with anyone as if they were already close friends. He has been described as kind, loving, a creative artist, an excellent storyteller and a dedicated researcher, and one friend called Noonan "a true RENAISSANCE MAN…ALL CAPS."[63]

A year after opening VPB, Noonan met with John Gallagher, and they decided to start a homebrewing club. The Green Mountain Mashers held their meetings at VPB, and Noonan was always keen to give advice and

Divin' Duck: Although extremely popular now, sour-mashed wheat ales weren't really in production in the United States in the 1990s. With a reddish hue and tartness from sour-mashing, an early menu described it as cider-like. A quote was included after the description—an homage to an old song called "Rye Whiskey": "If Champlain was a beer and I was a duck/I'd dive to the bottom and hardly ever come up."

Blackwatch IPA: The first commercially produced black IPA was brewed at VPB on December 3, 1994, by assistant brewer Glenn Walter under the auspices of Noonan. This beer solidified a brand-new beer style and would eventually be codified by the Beer Judge Certification Program (BJCP) in 2015. Because of the importance of this beer, the authors felt it only proper to dedicate an entire chapter to exploring it.

THE VERMONT PUB & BREWERY OF BURLINGTON

CRAFT BREWERS

BURLY IRISH ALE   Rich and malty, Burly is our house ale; a red Irish-style pub "session ale". It is mellow but brimming with character. It has a subtle Willamette hop character. We are one of a few breweries in the world that brews in the old Irish style. Never had an Irish ale? Similar to a Dos Equis Amber, but fresher, maltier and more flavorful.

Pint: $2.50   12oz Mug: $2.00

PESKY SARPENT LAGER   It's back! This unfiltered lager recalls a 19th century style of beer and brewing that has disappeared in this country. Because it is not filtered, the rich fermentation character has not been stripped from this beer; flavor has not been forsaken for clarity. Cellared for six weeks, it is brewed from brewers'. caramel, brown and Munich malts. It is sweetly-malty and mildly hopped for a subtle bitterness.

12oz Pilsner Glass: $2.00

GABRIEL SEDLMAYR OKTOBERFEST   Celebrate the harvest with our traditional Bavarian-style fest brew, named for the 19th century Munich brewer who preserved "Marzenbier" brewing. Our fest was cellared for two months. It is rich and malty, with the accent of Tettnang and Hallertau hops. On tap beginning September 22.

12oz Pilsener Glass: $2.00

DIVIN' DUCK SOURMASH WHEAT ALE   The wheat malt used in the brewing of this ale and its reddish color give it the appearance of cider; it is slightly sour, sweet, and rich for a quiet eve. Its aroma is intruiging.
"If Champlain was a beer, and I was a duck,
I'd dive to the bottom and hardly ever come up."

Pint: $2.50   12oz Mug: $2.00

SMOKED PORTER   Caramel malt smoked over hickory chips provides a smoky tang that perfectly compliments the sweet roastiness of our rich, black, full-bodied Old-English style Porter.

Pint: $2.50   12oz Mug: $2.00

JOE LIGHT - Wanna Lite? Drink our "Joe" Light. It's....light, it's reduced-alcohol, and it's absolutely fresh. Joe is "blended at the tap" with soda water in the Philadelphia-nouveau-tradition! With all due respect to brewer Joe Ortlieb:

12oz Pilsener Glass: $1.50

The Vermont Pub & Brewery of Burlington
P.O.Box 5477, 144 College Street, Burlington, VT 05402-5477

An early Vermont Pub & Brewery beer menu featuring beer styles that other brewers weren't even considering at the time. Note the sour-mash wheat ale. *Courtesy Vermont Pub & Brewery.*

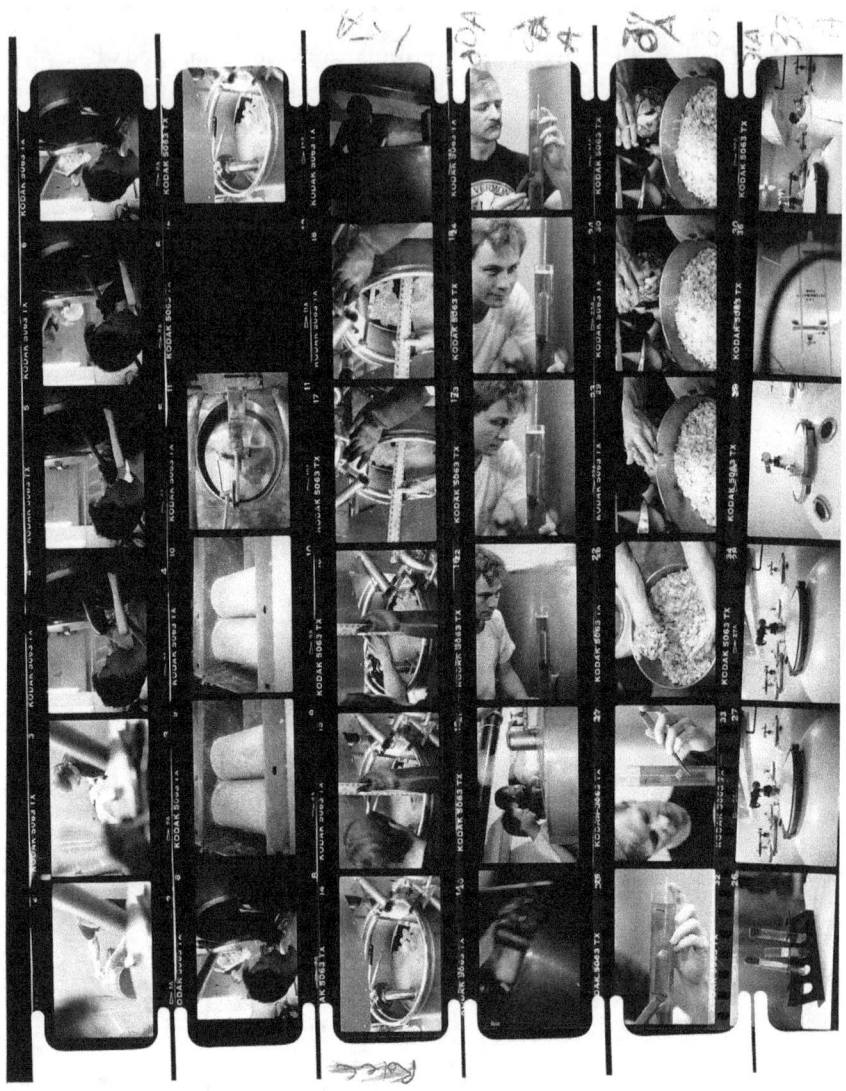

Above and opposite: Photo negatives from the very early days of Vermont Pub & Brewery. *Courtesy Vermont Pub & Brewery.*

suggest recipe formulations.[64] To add validity to the club, Noonan earned the title of National Beer Judge from the BJCP, which sets the standards for beer styles and judging requirements for competitions.

Noonan was a prolific writer, regularly contributing to many top brewing periodicals.[65] His dedication to honing his craft and to helping other brewers

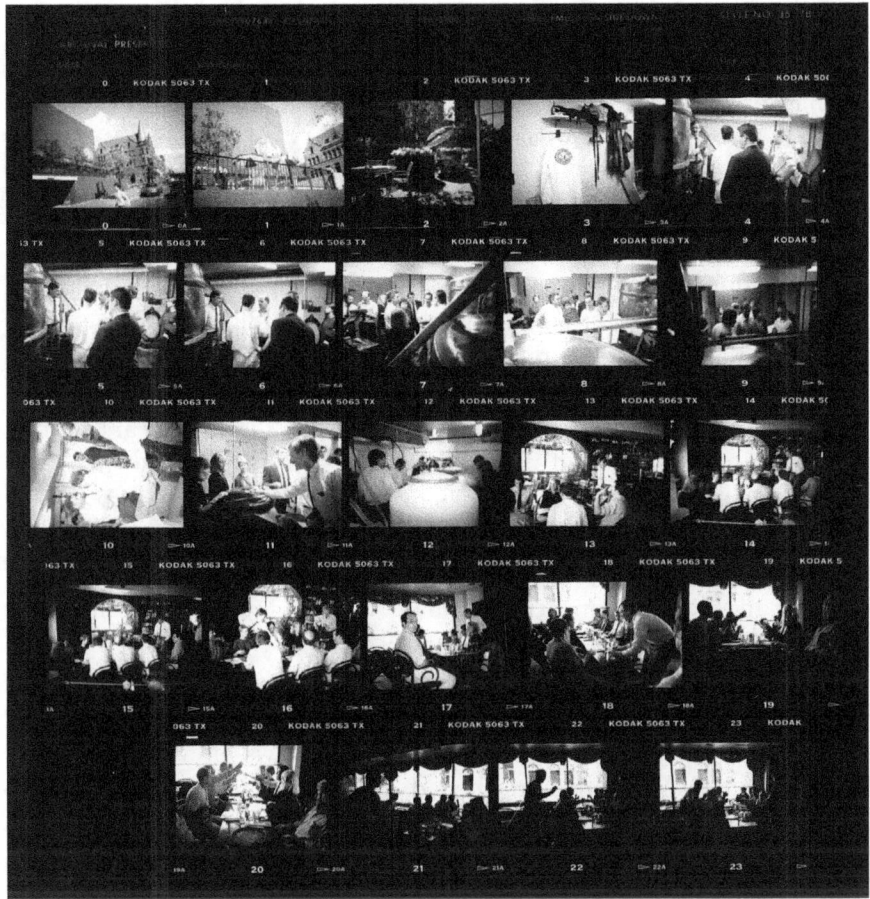

led him to write two more books: *Scotch Ale* (in 1993) and *The Seven Barrel Brewery Brewers' Handbook* (in 1996). In May 1995, Noonan cofounded the Vermont Brewers Association[66] with the goal of advancing quality brewing in the Green Mountains and promoting it beyond state borders.

The following year, the American Homebrewers Association honored Noonan with their Recognition Award, which "honors outstanding service to the community of homebrewers."[67] As Noonan's influence spread nationally, the Brewers Association awarded him the Achievement Award in 2004 and the Russell Schehrer Award for Innovation in Brewing in 2005.[68] While presenting the Schehrer Award, Steve Bradt, of Freestate Brewing Co., said of Noonan: "There are probably few brewers in the craft-brewing industry and even fewer homebrewers that haven't read one of Greg Noonan's books or articles on beer and brewing."[69]

NO NOONAN IS AN ISLAND[70]

Noonan wasn't just a talented brewer—he was also a generous teacher and collaborator. Many brewers got their start working at VPB. Glenn Walter went on to start Three Needs Taproom & Brewery. Jen and John Kimmich opened the world-famous Alchemist Pub & Brewery, now simply known as The Alchemist. But a start in brewing wasn't all that Noonan afforded to John Kimmich. He also supplied The Alchemist with their iconic yeast strain, Conan, a.k.a. VPB1188. And if one looks closely at the VPB logo, one will notice the rune that became The Alchemist's logo.

Dan Montanari came to VPB after brewing for four years at the Portsmouth Brewery and Smuttynose Brewing Company in New Hampshire and left in 2000. Later, Scott Gawitt became assistant brewer after graduating from the American Brewers Guild.[71] Sean Williams brewed for over three years at VPB before moving on to Otter Creek Brewing and then Boston Beer Company.[72] Other brewers have included J.D. Bowley, Andre Blais and Steve Miller.

The brewmaster's paddle was then passed to Russ FitzPatrick. FitzPatrick had worked at a homebrew shop, and when he decided to go pro, he attended the University of California, Davis, brewing program. He returned to Vermont and opened a brewery in Montpelier called Golden Dome, named after the golden top on the Vermont State House. Golden Dome was unique in that its beers were nitrogenated instead of carbonated, providing a rich, silky mouthfeel. When Golden Dome folded in 1999, FitzPatrick did a stint at the now-defunct Kross Brewing in Morrisville before Noonan offered him a job brewing at VPB.[73] FitzPatrick is the current brewmaster at VPB.

COLLABORATORS

In the spirit of exploration and innovation, VPB has become a hub for collaborations. In a 1999 interview, Noonan was asked about the strong sense of community between Vermont brewers. "I think it's because the craft-brewing movement was started by a lot of ex-flower children.... There's a lot of willingness to share information with everyone else. And not just in-state: We get calls from the West Coast, too. That's a complete turnaround from the old days when brewing was a very secretive, guild-run business."[74]

Many of the winning beers from the Green Mountain Mashers Competition have been scaled up and brewed at VPB, affording homebrewers access to a professional brewing system and exposure to the greater beer market.

To commemorate VPB's twenty-fifth anniversary in 2013, Polewacyk, FitzPatrick and Tommy Noonan (Greg's nephew) teamed up with Magic Hat to brew a special beer. The Magic Hat brewers, J.P. Williams and Justin McCarthy, brainstormed a mouth-puckering sour IPA called **Ol' Puckerface**. They brewed two test batches on the fourteen-barrel system at "the Pub," then scaled up the recipe to full production size at "the Hat." Williams and McCarthy left Magic Hat for other opportunities: Williams is now the brewmaster at von Trapp Brewing in Stowe, Vermont; McCarthy is a brewer at Zero Gravity Craft Brewery. Christopher Rockwood was named head brewer at Magic Hat and worked on the final iteration of the project. The name hit trademark roadblocks and was changed to **Steven Sour**. With this IPA brewed with passionfruit juice and fermented with the Conan yeast strain, VPB was once again ahead of its time—passionfruit would become a hot additive in the brewing world by 2017.[75]

REMEMBERING GREG

Greg Noonan passed away on October 11, 2009, at the age of 58. He had a short struggle with stage four lung cancer, and he chose to keep it private, so his death sent shock waves through the national brewing community. Beer writers like Stan Hieronymus wrote tributes and memorials for Noonan. Hieronymus wrote, "He was *always* smiling."[76] Paul Sayler, who first met Noonan while working as an apprentice at Catamount Brewing, wrote of his friend: "Greg mesmerized me…with his deep, FM-DJ voice and a stunning knowledge of beer history, he set me on the path that would lead me to become a brewpub brewer."[77]

Polewacyk took the helm at VPB and keeps Noonan's dream alive today. He accepts no praise, nor does he take any credit for the success of VPB. "It's all Greg Noonan…I'm the caretaker of his legacy. I do nothing more than that."[78]

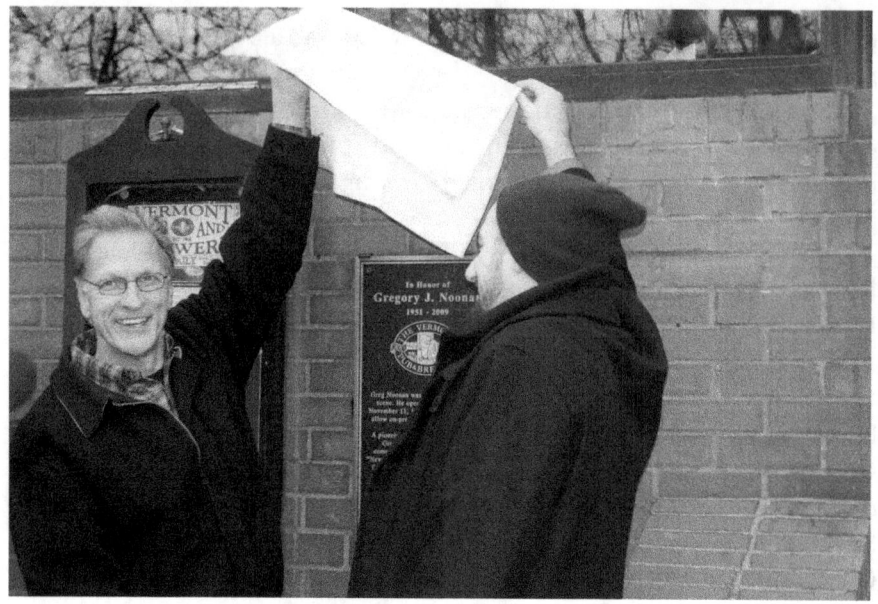

Steve Polewacyk (*left*) and Tommy Noonan unveiling the plaque dedicated in honor of Greg Noonan and his commitment to craft beer. *Courtesy Kurt Staudter.*

Since Noonan's passing in 2009, multiple breweries have released tribute beers in his honor:

VPB first brewed **Tulach Leis** (pronounced "toolah leez") in 2010. A Flanders-style red ale, it is fermented with Brettanomyces, resulting in a very dry, slightly sour wild ale with earthy barnyard notes mingling with fruity esters. The name is derived from County Cork in Ireland, where Noonan's ancestors lived, and was brewed by his friends at VPB in his memory.

Smuttynose Brewing brewed a black IPA and simply called it **Noonan**.

Berkshire Brewing Company released a bourbon barrel–aged Scotch ale called **Gude Greg's Wee Heavy Special Reserve**.

Paul Hale, brewmaster at Queen City Brewing, also brewed a Scotch ale to toast Noonan's memory: **Gregarious**.

Lawson's Finest Liquids brewed their own **Spruce Tip IPA**, and the bottle's label described the beer as a "unique brew inspired by Greg Noonan, the original VT spruce tip ale creator!"[79]

Hopfenstark Brewery (Quebec) brewed an Export stout dubbed **Greg**.

Dieu du Ciel! (Quebec) brewed an Imperial black IPA tribute beer called **Pioneer** in collaboration with Shaun e. Hill (Hill Farmstead).[80] Dieu du Ciel! would go on to formulate another black IPA called **Pénombre**, and the label depicted Noonan coaxing smoke out of a pint of beer as if he were a wizard performing alchemy.

In 2009, the Green Mountain Mashers renamed their annual competition the Greg Noonan Memorial Homebrew Competition.[81] The American Brewers Guild, founded by Steve Parkes and Christine McKeever, announced the Greg Noonan New England Brewer's Scholarship in 2013.[82]

"The industry is full of people who share what they know, but Greg really stood out," said Parkes. "He was a humble man who really, genuinely was willing to share with anybody who asked."

THE BLACK IPA

Outside of all the accolades and awards, Burlington can lay claim to something special in the brewing world. In 1994, a uniquely American beer style was born at Vermont Pub & Brewery: the black IPA. Although it wouldn't be recognized by the Beer Judge Certification Program (BJCP) as an official beer style until 2015, the black IPA spread across the United States and abroad.

The original black IPA, dubbed **Blackwatch**, is firmly documented as being brewed at Vermont Pub & Brewery on December 3, 1994. Owing to their thorough brew logs, brewmaster Greg Noonan and then–assistant brewer Glenn Walter, who now owns Three Needs Taproom, clearly produced the first commercial example of this style. Inspired by **Buffalo Bill's Brewery Alimony Ale** (IPA), Walter brewed the VPB variant of an IPA to reflect his feelings regarding his divorce proceedings, saying he wanted something "black and bitter."[83] The variant was a huge success, to say the least. Aggressively hopped with Cascade hops, it was significantly more bitter than most beers then on the market and inspired other brewers to attempt the style.

In 1995, John Kimmich, who would go on to found the Alchemist Pub & Brewery, brewed a batch of **Blackwatch** at Vermont Pub & Brewery. He would go on to create his own black IPA (**El Jefe**) at The Alchemist, which, in turn, inspired Shaun e. Hill to brew one (**Darkside Black IPA**) at the Shed Brewery in Stowe. Mitch Steele, then brewmaster for Stone Brewing Company in Escondido, California, tasted Darkside at a 2006 beer festival, and it inspired him to brew **Sublimely Self-Righteous Ale**, which went

Let the record show that Vermont Pub & Brewery produced the first commercially brewed black IPA, Blackwatch, on December 3, 1994. *Courtesy Vermont Pub & Brewery.*

on to be considered the benchmark of the style.[84] The style also popped up in Oregon around 1996 at Rogue Ales in Newport.

When Noonan passed away in 2009, brewers issued tribute beers in his honor. In 2013, Smuttynose Brewing Company of New Hampshire brewed a black IPA and simply called it "**Noonan**," adorning the label with Greg's likeness. *Wine Enthusiast* magazine gave the beer ninety points (out of one hundred), saying it showed "good interplay between the roasty, chocolaty flavors and the hop-derived bitterness, ending harmoniously on notes of bitter cocoa."[85]

As the style moved across the country and took root on the West Coast, an ongoing debate started as to whether it should be called a black IPA or a Cascadian dark ale. Brewers located in the Pacific Northwest argued that the beer style relies heavily on the Cascade hops grown in their region and should therefore be called Cascadian dark ale. But after much debate, black IPA eventually won out as the "official" BJCP style name.

In the brewing world, it is generally considered important to look at how the Beer Judge Certification Program has defined a beer style. The black IPA could have been relegated to a catch-all category called "American Black Ale," but in 2015, the BJCP added "Black IPA" as its own substyle in the IPA category.

As these style guidelines evolve, one thing remains the same: black IPA is one of the very few American styles. There is a point to make, though, since one would argue that "American" appears before many BJCP beer style guideline names. Black IPA was developed by changing malts and hops to create a unique style. Many of the "American" hyphenated styles are defined by brewers simply adding more hop character and achieving a higher IBU (international bittering unit) value than those of the beer style's predecessors. For example, American IPAs typically score higher bitterness levels than English IPAs. Other uniquely American styles are the Cream Ale, Kentucky Common and California Common. An argument can be made that these are all variants of European beers; the California Common, or steam beer, has its roots in the Bavarian Forest with Dampfbier, but the differences are great enough that it warrants a separate category.

While **Blackwatch** was a totally new riff on the IPA, the black IPA has one long-lost connection to England. Burton ales, a style of ale brewed in the Burton region of England, were known in the nineteenth century for being very alcoholic, bitter ales that reached toward 11 percent ABV. The region was also known for producing IPAs that traveled well on ships. In an 1888 brewing manual from London, there was a mention of a rather interesting pale ale produced in the area:

> *To begin with, then, it is not customary to employ saline waters, or, in other words, if such water be employed the black beer produced is deficient in*

that roundness and fullness of palate taste that is considered so necessary a feature, while I can example this by referring to the black beer produced at Burton, which has been universally described as a mere black pale ale – i.e., though black in colour, its palate taste reminds one very strongly of the pale beers produced by Burton firms. It will be quite understood that I am not decrying this article; it may and does suit many palate tastes, and is thought a great deal of on the Continent, but at the same time it differs very widely from the accepted standard quality of a black beer as specified.[86]

This beer would theoretically have been a very hoppy take on a black beer and more akin to a dry porter style. The manual from which this passage was taken was published exactly a century before Vermont Pub & Brewery opened their doors in Burlington.

MAGIC HAT BREWING COMPANY

5 BARTLETT BAY ROAD, SOUTH BURLINGTON

T he story of Magic Hat Brewing Company is a tale that began with a self-proclaimed "serial entrepreneur" and a homebrewer. Alan Newman and Bob Johnson certainly pulled a rabbit out of a hat when they founded Magic Hat in early 1994, starting with a name that everyone hated and growing the company into a multimillion-dollar brewery.[87]

UNLIKELY MAGICIANS

Newman wasn't a beer guy. He was an idea guy—a marketing savant. An automobile fanatic born in Brooklyn, Newman declared himself a "hippie" in 1966, and after a bevy of jobs, Newman forged his own way. He founded Seventh Generation, a natural household products company, but was ousted. Looking for his next adventure, Newman bumped into his friend Bob Johnson while walking on Burlington's Church Street.

Johnson was a beer guy. He brewed his first batch in his bathtub in 1978. He honed his homebrewing skills in his spare time while working as a cheesemaker at Shelburne Farms and eventually decided to open his own brewery. When the two ran into each other that fateful day, Newman expressed interest in opening a brewery in Burlington with Johnson.[88]

What's in a Name?

Any good story needs a name, and it was a process to settle on the name for their new brewery. The duo knew what they *didn't* want: no geographical references, nothing about Vermont and definitely no using their own names. They wanted a name that meant nothing specific at all. That way, they could be as creative as possible with their branding. Johnson proposed Magic Hat, and although Newman liked it, everyone they asked said, "it sucks."[89] That only spurred them on, and Magic Hat Brewing Company was about to appear.

Early Days

The first batch of beer the company released was **Magic Hat Ale**, a robust Irish-style red ale, which Johnson brewed at Federal Jack's Brewpub in Kennebunkport, Maine.[90] The beer was well received in Burlington, and plans were set to build a brewery on Flynn Avenue. A fifteen-barrel brewhouse was installed, but a bottling line was too expensive. The Flynn Avenue brewery would only produce draft beer, while they contract-brewed beer at Kennebunkport Brewing and had it bottled at Shipyard Brewing.[91]

Magic Hat's signature style was established early by Johnson's choice of yeast strain and fermentation techniques. Unlike most breweries, where fermentation occurs in closed vessels, Johnson chose to utilize open fermentation, which can be a little tricky. He selected an English ale yeast called Ringwood, which beer writer John Holl has noted "can give signature fruit flavors, but is notoriously difficult to work with."[92] Johnson and his brewing team mastered Ringwood and have used it for many of Magic Hat's biggest hits.

Mardi Gras Parade

It doesn't seem to matter that Mardi Gras occurs in the middle of Vermont's winter; Burlingtonians love a good excuse for a party—especially when the party benefits nonprofits.

In 1996, Stacey Steinmetz, Magic Hat's first employee, proposed the idea of hosting a Mardi Gras parade in Burlington. It wasn't an immediate hit

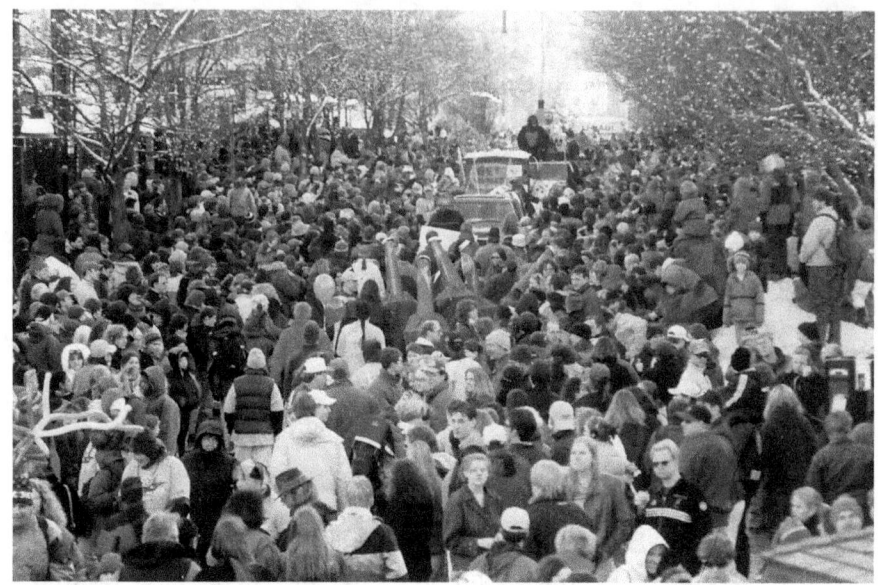

Crowds surrounded the Vermont Pub & Brewery team—dressed up as beer bottles (*center*)—at the Magic Hat Mardi Gras Parade. *Courtesy Vermont Pub & Brewery.*

with the top brass. Years later, Newman admitted that although he often gets credit for the parade, he actually said that it was "the stupidest idea ever!"[93] But Newman was glad that Steinmetz pushed forward on it and admitted that she was right—it ended up being a huge success.

The first parade was held in 1996 and featured fifteen floats designed by area businesses, and about 1,500 people attended. Proceeds from the parade benefitted Vermont Cares, an AIDS awareness nonprofit.[94] The 2018 parade attracted roughly 25,000 attendees and raised enough money for the Vermont Foodbank to supply 35,000 meals to Vermonters in need. Big-name music acts are brought in during Mardi Gras weekend, and notable bands have included Gogol Bordello, Fishbone, moe., Lotus, Reel Big Fish and Rubblebucket.[95] In 2019, Magic Hat retired the parade, hosting a block party at Burlington's Church Street Marketplace instead. Since they began, the Magic Hat Mardi Gras celebrations (including the 2019 block party) have raised over $260,000 for Vermont nonprofits.

MOVING ON AND OVER

In an effort to increase production capacity, Magic Hat purchased a shuttered Key West brewery in 2002 and shipped all of its equipment up to Vermont.[96]

As the brewery grew, Johnson left Magic Hat in 2003. (Newman claims they had to push him out but says they've stayed friends.) Magic Hat had grown to the point of needing to brew three shifts a day and project many years into the future for their raw material needs. Johnson was "more an artist than a businessperson," according to Newman. Johnson accepted a settlement[97] and moved to Portland, Maine, where he cofounded Scratch Baking Co.

Now, Newman found himself with a problem; when a brewery grows at a breakneck speed, it will eventually run out of capacity and need to expand. After securing loans, buildout started at what would become Magic Hat's final home on Bartlett Bay Road in South Burlington.[98] As a former lumberyard,[99] the property gave Newman a blank canvas to shape the future of the Magic Hat experience.

The new, thirty-six-thousand-square-foot brewery, complete with a bottling line and visitor center (The Artifactory), opened in September 2007.[100] Famed designer and builder Russ Bennett, of NorthLand Visual Design & Construction, designed the "steampunk chic" Artifactory.[101] Bennett would later work on another Burlington brewery—Foam Brewers.

Rapid growth continued, and Magic Hat expanded distribution from just the Burlington area, in 1994, to selling in over forty states by 2018. Head brewer Christopher Rockwood estimated that they would brew about eighty-one thousand barrels of beer by the end of 2018.[102]

INNOVATION AND EXPANSION

In June 2007, Newman and his team decided to push the envelope even further, creating a subsidiary: Orlio Organic Beer Company. At the time, Wolaver's Organic was the only Vermont brewery operating in the organic craft beer arena. Orlio had a completely different vibe than Wolaver's. Its packaging was minimalist: a black background and a ring of color denoting which beer was in the bottle. Three flavors made it to market: **Common Ale**, an IPA and a Schwarzbier.

In 2008, the **Common Ale** took a Gold Medal in the World Beer Cup. But Orlio was ahead of its time, and when the project ended in 2009, the company cited a lack of sales.

Another avant-garde innovation incubated at Magic Hat was the country's first anaerobic digester designed to process brewing waste products. In 2010, Eric Fitch, founder of PurposeEnergy, worked with Magic Hat to design and install a waste treatment plant.[103] The digester is capable of cleaning and recycling 360,000 gallons of water per year and producing roughly 30 to 40 percent of the brewery's energy needs via naturally produced methane gas. The digester is also capable of removing phosphorus from the wastewater.[104]

Purchases and Sales

Starting in 2008, Magic Hat had major structural changes about every two years. In 2008, Independent Brewers United (IBU), Magic Hat's parent company, purchased West Coast Pyramid Breweries for somewhere between $25.2 million and $35 million.[105] This turned out to be disastrous, according to Newman,[106] and would lead to the sale of Magic Hat.

North American Breweries (NAB), which managed Genesee, Dundee and Labatt USA and was itself owned by a New York City capital firm, purchased IBU in 2010.[107] After a couple more years and an exodus of brewers, NAB itself went up for sale. In October 2012, KPS Capital Partners, owners of NAB, announced they would sell their entire portfolio for $338 million in cash to Cerveceria Costa Rica, a subsidiary of Florida Ice and Farm Company.[108]

The most recent big change came early in May 2017. This time, it wasn't a sale or a purchase; rather, the Artifactory was redesigned to welcome visitors with cheeseboards, sandwiches and other tasty treats in addition to full pints of beer.[109]

Celebrating the Art

The original name for the brewery when it was on Flynn Avenue was Magic Hat Brewing Company and Performing Arts Center. Newman envisioned a brewery that was totally unique compared to the (four) other Vermont breweries in existence at the time. Although the "Performing Arts Center" was dropped, Magic Hat remained dedicated to the arts, hosting concerts at the brewery and brewing an annual **ArtHop Ale** to benefit Burlington's South End Arts and Business Association. In both 2016 and 2017, Magic Hat was recognized by Americans for the Arts, which highlights "businesses of all sizes for their exceptional involvement with the arts that enriches the workplace, education, and the community."[110]

FAMOUS BREWS

With a catalogue of over one hundred beers,[111] it's hard to whittle down a list of which beers to include, but here are some of the most noteworthy Magic Hat brews:

Bob's First Ale: The first commercially brewed beer Magic Hat released was this Irish-style red ale with a prominent malt character. It garnered a following in Burlington bars and pubs and has stood the test of time, eventually becoming a winter seasonal.

#9: Designed to be a lighter offering (as the first summer seasonal), #9 was a "not quite pale ale" infused with apricot essence. In his memoir, Newman said that everyone was floored by the rapid sales rate, which showed continued increases for fifteen consecutive years.[112] While Newman pointed to the name of the beer as contributing to its success, Paul Sayler of Zero Gravity Craft Brewery cited the Ringwood yeast. Ringwood "certainly was very fruity. And maybe that's why #9 was such an immediate hit. [It] totally made sense because the yeast was already ready for apricot essence to join it. You know, it already had various points where that would just mesh into the overall ester profile."[113] This would become Magic Hat's flagship beer.

Blind Faith: The first Magic Hat IPA, Blind Faith was released simultaneously with #9. Originally below forty IBUs, it has crept up to fifty IBUs in recent batches. The beer is decidedly on the maltier side of the IPA category and has come and gone from regular production over the years.

Feast of Fools: This raspberry-infused American stout went from being a thank-you gift to the company's best retailers to a winter seasonal. Originally released in black champagne-style bottles, Feast of Fools moved into sixteen-ounce cans in 2017. Feast of Fools is part of the Humdinger Series, which helps raise funds for nonprofits, and in 2017, Vermont Foodbank was the beneficiary.[114]

Chaotic Chemistry: This barleywine ale spent two years in bourbon barrels and then slumbered for another year in stainless steel.[115] It was brewed to such a high ABV that it was only available in Vermont liquor outlets until laws changed in 2008.[116] After 2008, beers with up to 16 percent ABV could be sold in private retail stores. Also part of the Humdinger Series, Chaotic Chemistry was packaged in iconic black champagne-style bottles.

Heart of Darkness: In 1995, the late beer writer Michael Jackson tipped his hat to a mysterious stout called Heart of Darkness, noting "the density of its blackness."[117] A clear reference to Joseph Conrad's book, Heart of Darkness was described by the brewery as being "filled with the howling of black dogs that haunt the long-forgotten shadows of the human soul." This beer also received praise from Garrett Oliver (Brooklyn Brewery) and beer writer Fred Eckhardt, who declared it "flawless."

Ale of The Living Dead: A one-time release in clear bottles, this was brewed without any hops and was chock full of garlic. "Why beer god?! Why?!" wrote Todd Alström, cofounder of BeerAdvocate.com, in a review.[118] Former head brewer Matt Cohen marked the beer down as the worst beer he's ever brewed. "Within a day, all the beer got recalled because it was just so damn undrinkable. We destroyed most of it, but we kept a couple of cases, and we used to do bets. If you lost a bet, you'd have to drink, like, two of those bottles."[119] Magic Hat did receive some kudos for their creativity and for pushing boundaries, but this beer was immediately retired.

Wacko: In 2009, Wacko, a sessionable, pink-hued beer, hit the market. The pink coloring was thanks to the addition of beet sugar and beet juice. It was created based on the popularity of a test batch called "Kerouac," which was brewed around 2007. Named, of course, for beatnik author Jack Kerouac, the recipe evolved into Wacko and was introduced as a summer seasonal. It has since been retired.

Big Hundo: This imperial IPA was originally designed around Vermont-grown hops and named for its 100 IBUs. It is now produced in partnership with The Hops Project, which "supports Vermonters looking to grow hops," according to Heather Darby of the University of Vermont Agriculture Extension Program.[120]

BELIEVING IN MAGIC

Paul Sayler of Zero Gravity Craft Brewery believes in the magic and wants others to as well. He sees Magic Hat as a spring of innovation in Vermont's brewing culture. "I always was a big booster of the place and loved the beers and loved the people that worked there. Magic Hat needs to be considered as one of the really significant pillars of Vermont brewing. But unfortunately, because of the sale and that exodus of brewers after the sale that sometimes we might forget how important Johnson and Newman and all the brewing staff over the years were to the Vermont scene. And still are. They brewed great beer and occasionally amazing beer over the years."

MAGIC HAT ALUMNI

Dozens of brewers got their start at Magic Hat, and many of them have gone on to work for other Vermont breweries or to start their own. Matt Cohen founded Fiddlehead Brewing in Shelburne, Vermont, and John Ravell would eventually join him as general manager. Mike Gerhart became brewmaster at Otter Creek Brewing in Middlebury and then COO for Hill Farmstead in Greensboro. Todd Haire and Dani Casey would team up with three other brewers to start Foam Brewers in Burlington. Haire would also establish House of Fermentology with Bill Mares. J.P. Williams left Magic Hat and became brewmaster at von Trapp Brewing in Stowe. Scott Martin, who also brewed at Three Needs, moved to Canada and brewed for McAuslan Brewing, Les 3 Brasseurs and Townsite Brewing. Justin McCarthy went to Zero Gravity Craft Brewery. Lillian MacNamara helped to open Hop'n Moose Brewing (now Rutland Beer Works) and eventually became head brewer at Queen City Brewery in Burlington. And Mark Babson opened his own brewery, River Roost Brewing, in White River Junction.

Alan Newman is himself an alumnus of Magic Hat. He left in 2010 and joined forces with Stacey Steinmetz to start Alchemy & Science, a craft beer brand incubator for Boston Beer Company.[121] When Newman's contract expired at the end of 2016, he left the beer industry and purchased a stake in Higher Ground, the Burlington area's premier music venue.[122]

THREE NEEDS TAPROOM & BREWERY

185 PEARL STREET, BURLINGTON

*"You couldn't pin the place down. It had its own personality,
some of which was hidden from view."*
—Paul Sayler

There once was a small brewery called Three Needs Taproom & Brewery, on College Street, where locals gathered to watch reruns of *The Simpsons* and drink pints of beer brewed in the basement. A small bar, it was heralded by many locals as one of the most unique spots in Burlington to chat with strangers. And for seventeen years, Three Needs served patrons a rotating bevy of ales and lagers at that location before moving to a significantly larger location a few blocks up on Pearl Street in 2002.

THE ROOKIE SEASON

When asked what year Three Needs opened, founder Glenn Walter replied, "1995…Derek Jeter's rookie season."[123] This was also Walter's rookie season. After brewing with Greg Noonan at Vermont Pub & Brewery for many years, Walter felt it was time to go pro on his own. Noonan wasn't particularly pleased with this decision at the time, and Walter had to wait out a one-year noncompete agreement before he could start his own brewery.

While ripping the calendar pages off day by day, Walter opened a bar at 207 College Street with the intent to eventually outfit the basement with a brewery. When Walter's noncompete agreement expired, he brewed his first beer for his new operation: a German-style Sticke Alt.

Before we go any further, one might ask: "What are these so-called 'three needs'?" The logo and the name came to Walter in a dream, and he has protected the mystery of its meaning, asking bar patrons to write down their own "three needs" in a volume of notebooks kept at the bar.[124]

PLAYING THE BEER BY EAR

For some, brewing is a science. For others, it's more like jazz, with lots of improvisation. Good jazz players know how to artfully break the rules. It's often the same with brewers, and Walter always liked to brew by ear, so to speak.

"That's the one reason I always chose owning a brewpub instead of a microbrewery...I always wanted the evolution....Some people are really into, like, perfecting that one thing and narrowing that pencil point down and get it better and better and better. For me, it was just always the evolution of trying to make it better, and maybe one batch wasn't as good as the last batch. But it was always a one-step-forward progress for me."

While most other Vermont breweries were focusing on amber ales, pale ales and IPAs, Walter preferred to brew beer styles that weren't readily available. He loved to brew Continental-style pale lagers and Belgian-style ales, focusing his attention on proper yeast selection instead of other ingredients. Walter was also quite fond of dark ales. **Chocolate Thunder Porter** was one of the only beers that one could always find on tap while visiting Three Needs.

Brewed with chocolate malt and named for locally famous bouncer Mikey van Gulden, Chocolate Thunder Porter became a Burlington standard. But despite the beer's permanent spot on the draft tower, Walter allowed his brewers to make subtle changes from batch to batch in keeping with his belief in constant evolution.[125]

"DUFF HOUR"

When he first opened the bar, Walter figured opening at noon would bring in locals. But what he found was that the noon to four o'clock crowd really bummed him out. He decided to switch the opening time to 4:00 p.m.

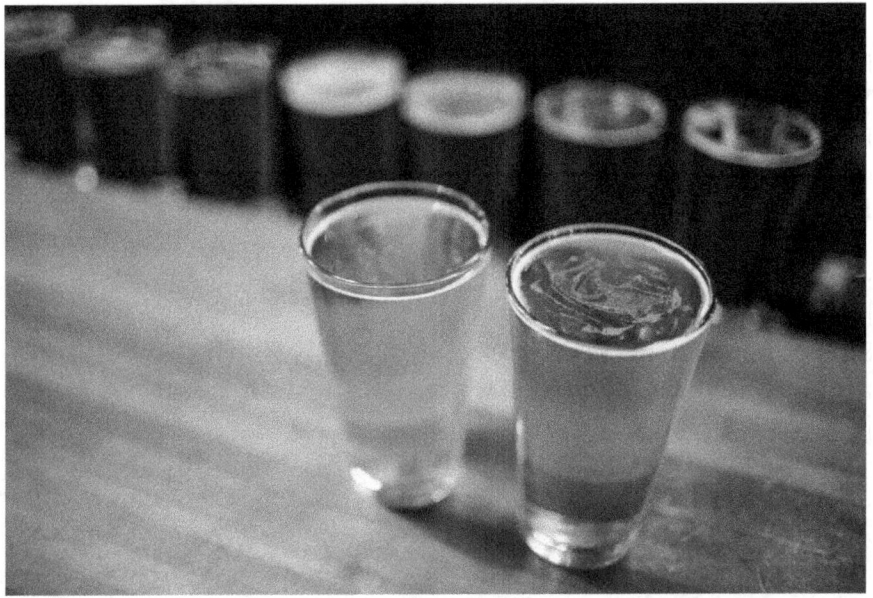

Pints at the ready for "Duff Hour" at Three Needs Taproom & Brewery in 2017. *Courtesy Paul Sarne.*

and, in an effort to draw a jollier demographic, he would play reruns of *The Simpsons* on the TV with the volume turned up loud. To encourage even more business early in the afternoon, he decided that every day, when they opened the doors, he would tap one keg of beer and sell pints for one dollar until the keg kicked. He named this "Duff Hour" after the fictitious Duff Beer in the world of *The Simpsons*. The bar started to collect *Simpsons* memorabilia, adding to the funky vibe, and Duff Hour has been a Burlington staple ever since.

THE BREWERS

Once the three-and-a-half-barrel brewhouse was operational, Walter quickly realized that it was one thing to be a full-time brewer and another thing to be a bar owner and operator *and* a full-time brewer. He had to fill up a menagerie of seven-barrel and three-and-a-half-barrel fermenters, some of which were modified old Grundy tanks. Walter needed help; he brought in Dan Lipke to brew with him.

Lipke had brewed his first batch of homebrew in 1990, and he was immediately hooked. Coincidentally, Lipke had also started his career at Vermont Pub & Brewery.[126] By 1999, Lipke and Walter had brewed over forty-two different styles of ale and lager.[127] Lipke went on to brew for the original Trout River Brewing in Lyndonville, Vermont, and then Mercury Brewing and Clown Shoes Beer in Massachusetts.[128]

In 2001, Jeffrey Hughes took over as head brewer and was focused on mixing English-style ales into the lineup of regular house beers. Hughes left a year later and spent eight years brewing at the Original Saratoga Springs Brewpub. He bounced around in the liquor industry for a while and eventually landed a sales job with Country Malt Group, a wholesaler of high-quality brewing ingredients.[129]

James "JT" Tierney also started at Three Needs in 2001, first working as a manager and eventually taking over brewing. Famed beer writer Andy Crouch wrote that Tierney brewed over 140 different kinds of beer annually.[130] After five years and a handful of award-winning recipes, Tierney did a short stint at Portsmouth Brewery in New Hampshire and then went into a career in the biomedical industry.[131]

Then there was an unemployed beekeeper named Scott Martin who wanted to break into a job in either farming, baking bread or brewing beer. Martin was a regular at Three Needs, and both Walter and Tierney gave him a tip that Magic Hat Brewing Company was hiring. After brewing full-time at Magic Hat and helping out part-time with brews at Three Needs, Martin was named brewer at Three Needs and worked there from 2009 to 2010.

During his tenure at Three Needs, Martin followed Walter's ethos of having an ever-changing beer repertoire. "At Three Needs, the only mandate from Glenn was to have one dark beer on at all times," according to Martin. "Coming from a structured environment like Magic Hat, the complete freedom was almost overwhelming; I didn't know what to do with myself. Some days I would go in with no plan, only knowing I had to brew."[132]

With a penchant for brewing with Vermont ingredients, Martin crafted a Bière de Miel, or honey saison, called **Hexagon**, which was well received by bar patrons and industry veterans alike. Martin recalled that he fermented it with the Dupont yeast strain, and when that stalled out, as it is wont to do, he pitched in Brettanomyces Claussenii and let it sit in a conditioning tank for two years, where it likely picked up some ambient Pediococcus. Martin eventually moved to Canada with his wife and has since brewed for Quebec's McAuslan Brewing and Les 3 Brasseurs and is currently brewing for Townsite Brewing in British Columbia.

As the attentive reader may have already noted, Burlington has been home to a number of talented homebrewers who have gone pro. While working as a bartender at the Farmhouse Tap & Grill, Paul Fletcher picked up a part-time gig brewing at Three Needs when Martin left. Fletcher was Three Needs' last brewer at the College Street location and went on to brew for Migration Brewing in Portland, Oregon.

A PEARL OF A LOCATION

When the lease for the building expired, Walter needed to move his business. He decided to sell his Sadie Katz Delicatessen and use the capital to purchase a building so he wouldn't have to worry about a lease ever again. The building that he purchased as the new home for Three Needs has a long history and a funky personality.

Located at 185 Pearl Street, the building is a large, meandering structure that previously housed a number of different businesses. Originally used as a carriage manufacturing and repair shop in the 1860s and then later as a residence for Dr. William Root, it eventually housed a high-end women's tailoring shop.[133]

In the mid-1970s, an eccentric woodworker took over the property and spent over three years completely remodeling the interior. Brian Fox, who formerly owned two hip counterculture clothing stores with his wife,[134] poured his heart and soul into fabricating a complex and beautiful interior for his restaurant called the Déjà Vu Café. Fox installed thousands of board feet of rock maple and other hardwoods, carefully hiding screws with countersunk wood plugs. Handcrafted brass chandeliers, stained glass, interactive decorative fixtures—Fox had a vision and couldn't stop until he was satisfied with the quality of his work.[135] When he was finally done, the Déjà Vu Café opened and was very successful for over a decade.

In 1984, Fox sold the Déjà Vu Café to the owners of Carbur's Restaurant and moved to Charlottesville, Virginia, with his family. But the new owners, Carl L. Capra and Burr Vail, soon had a falling out, and when Capra committed suicide in 1985, his estate sold the Déjà Vu Café to Pat and Robert Fuller.[136]

When the Fullers took over, they made another quirky addition to the building. In a small one-story building that was added to the property in the early 1900s, the Fullers worked with Burlington architect Brad Rabinowitz to recreate the so-called Frank Lloyd Wright Room at the New York Metropolitan Museum. The Frank Lloyd Wright Room at Déjà Vu garnered some national press when it was featured in *Metropolitan Home* magazine in 1987.[137] It remains intact today.

The Fullers sold the business in 1992[138] to Wanvadi "Jip" Jotikasthira and her husband, Daryl Campney, who changed the business to Parima Thai Restaurant. That's not all they changed. Campney had a stained-glass window crafted of his wife sitting in the nude by a stream and installed it high up on a brick wall near the vaulted ceiling. When Walter bought the property, he kept the piece in place, and she looks out over the pool tables as the matron saint of the bar.

CONTRACT BREWING

Walter had to make a tough decision when he moved to the Pearl Street location. The cost of outfitting a brewery was astronomical, so he pursued another avenue: contract brewing. After asking around, he discovered that Queen City Brewery on Pine Street already had his preferred ale yeast strain. In 2015, Walter brewed a Belgian-style witbier, a Schwarzbier and **Chocolate Thunder Porter** at Queen City Brewery[139] and had the beer shipped to the new Three Needs. Walter also teamed up with Drop-In Brewing in Middlebury, Vermont, to brew his Belgian-style beers, including the **Three Needs Tripel**. The house beers were finally back on tap at Three Needs.

THE MOST INTERESTING CORNER IN TOWN

The intersection of Pearl and Winooski Streets has its own hashtag: #TheMostInterestingCornerInTown. In addition to being home to Three Needs, this corner sports a collective of Burlington's quirkiest watering holes, such as the Other Place ("The OP" to locals); Radio Bean; a lamp shop that doubles as a music venue; and ¡Duino! (Duende), which serves international street foods late into the evening. There are also four Asian eateries, a CBD and hemp goods shop, a tattoo parlor and an emergency medical location for over-intoxicated individuals. One can imagine why this corner has earned its hashtag.

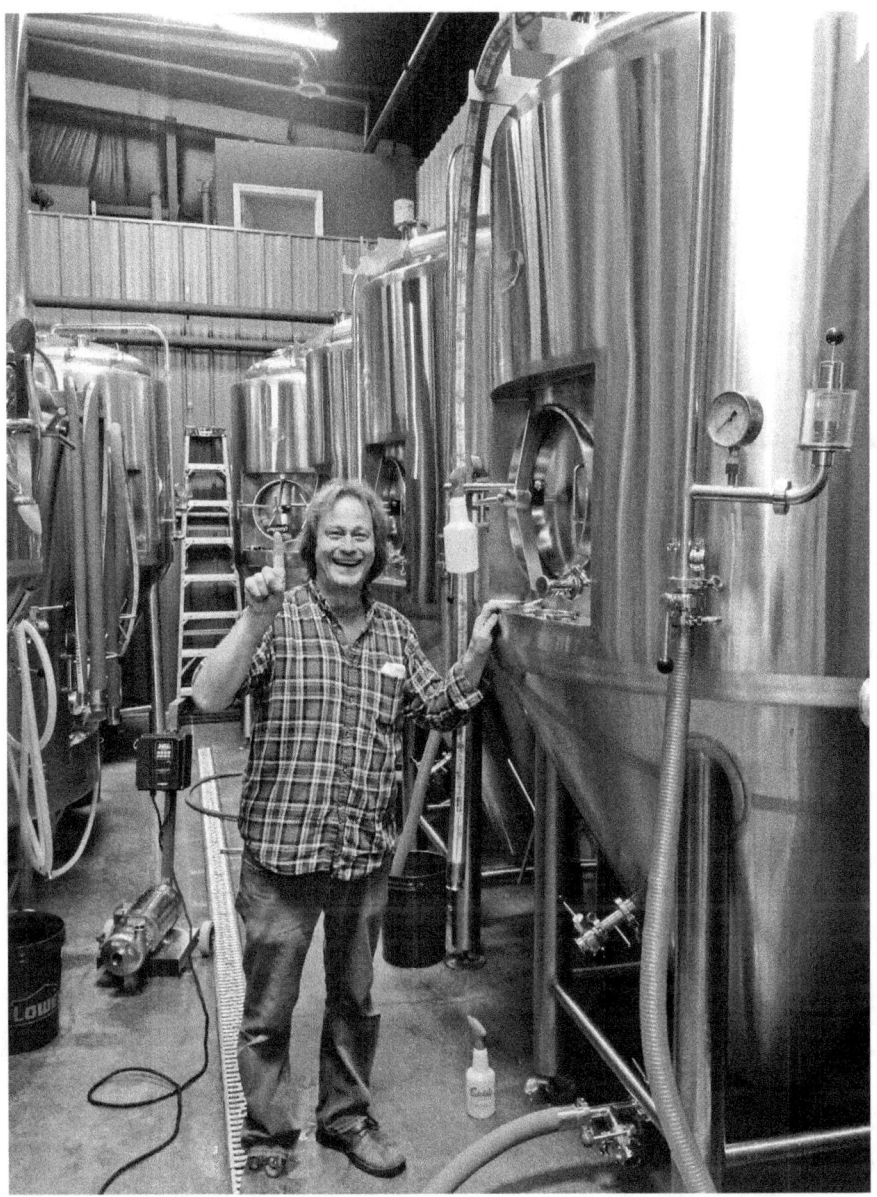

Glenn Walter brewing the first contract batch of his Three Needs Schwarzbier at Queen City Brewery. *Courtesy Glenn Walter.*

When Walter moved to Pearl Street, his goal was "to bring all of Main Street and all of College Street to this side of town." He believes in the adage "a rising tide floats all boats" and considers these other businesses to be "teammates, not adversaries." In addition to having a great spot to drink beer, play pool and converse with friends, Walter had another idea to keep patrons on his side of town.

PIZZA AND BEER

After about a year at the new location, Walter needed to add a food option. Burlington bars are required to offer some sort of food options to patrons and he joked that at the College Street location the health inspector considered a coin-operated candy machine to be "good enough."

So what sort of bar food would work at the new location and also be profitable? After contemplating the take-out window concept at Burlington's world-famous Nectar's Restaurant, Walter decided to emulate that model. He could serve patrons already inside his bar and also offer a snack to those walking by who were headed home or to the next bar. After scrapping a hot dog menu, he reconsidered. "[American] Flatbread does good with beer and pizza, so, yeah, I'll do pizza." Walter managed a pizza shop on Martha's Vineyard, so he felt confident with that business model,[140] and locals have embraced his pies.

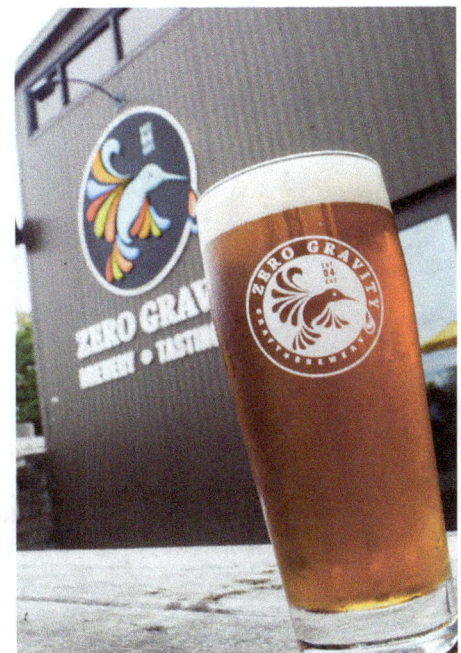

Right: A fresh pint of lager at Zero Gravity Craft Brewery's Pine Street location. *Courtesy Paul Sarne.*

Below, left to right: Justin McCarthy, Margaret Leddy and Chris Costello raising a beer after work at Zero Gravity Craft Brewery on Pine Street. *Courtesy Paul Sarne.*

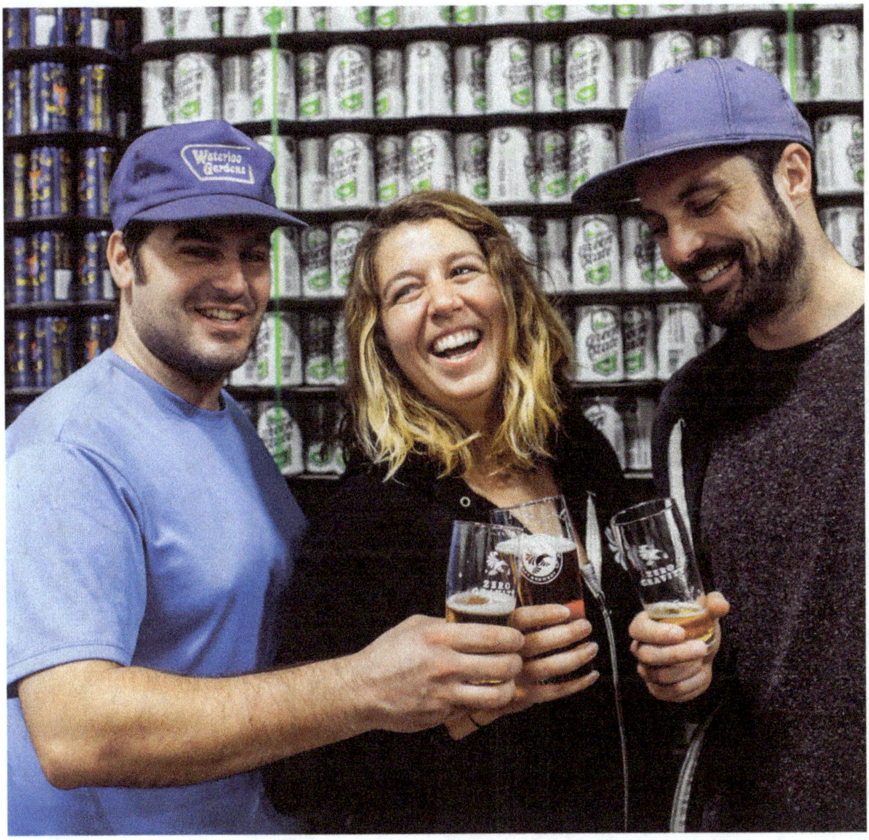

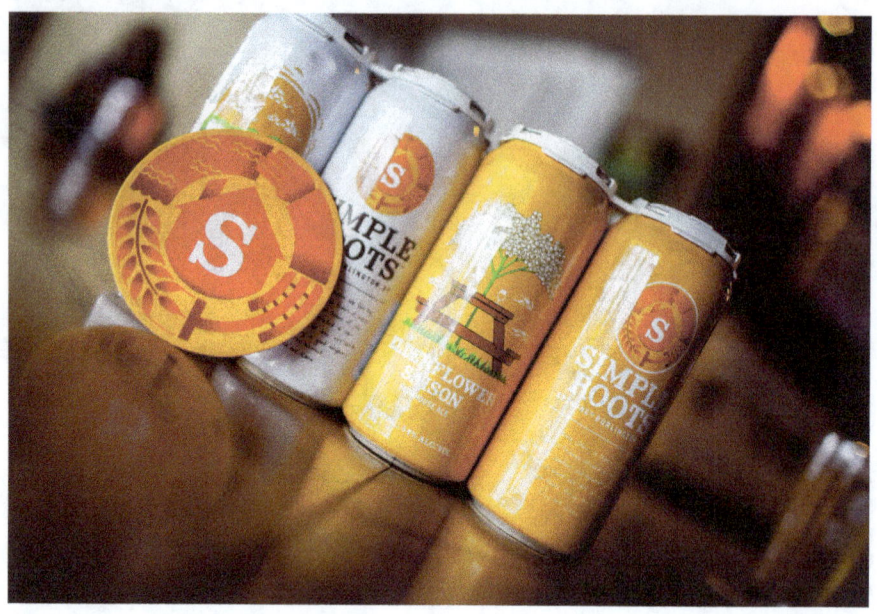

The branding and artwork for Simple Roots Brewing is meant to convey the feeling of Vermont foliage. The design work was done by Jeff Holmes. *Courtesy Paul Sarne.*

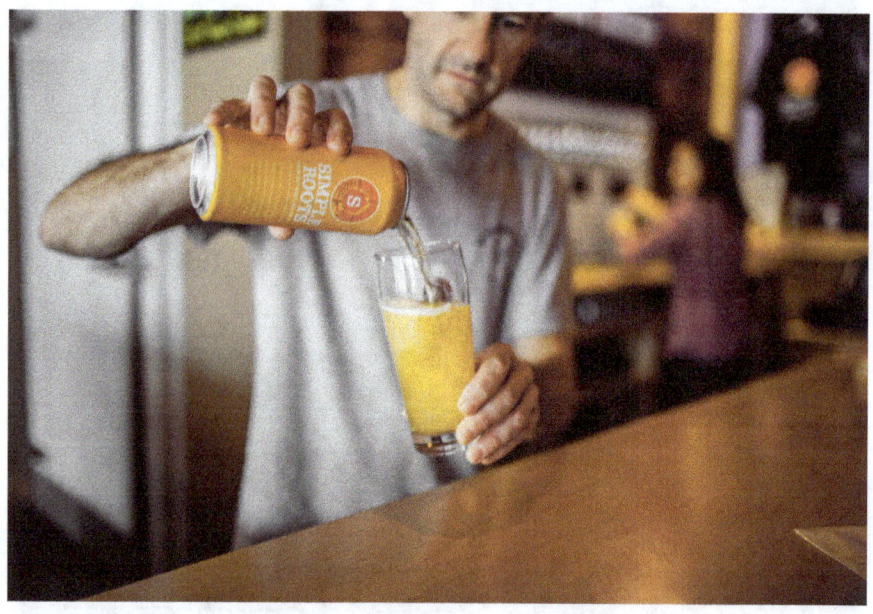

Dan Ukolowicz pouring a can of Simple Roots Elderflower Saison while his partner Karen Pawlusiac pulls a draft for a customer in their Old North End taproom. *Courtesy Paul Sarne.*

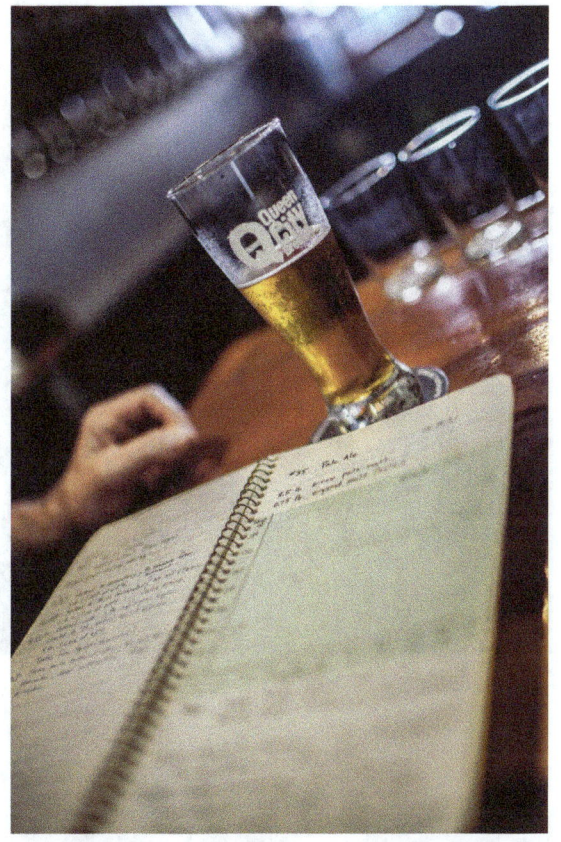

Right: Paul Hale, now brewmaster of Queen City Brewery, kept diligent scientific notes while he was home brewing. This is one of his first notebooks—complete with charts and graphs. *Courtesy Paul Sarne.*

Below: The Burlington craft beer and music scenes directly support each other. Jason "Metal Jay" Held sports his Burlington Record Plant shirt while managing the bar at Queen City Brewery. *Courtesy Paul Sarne.*

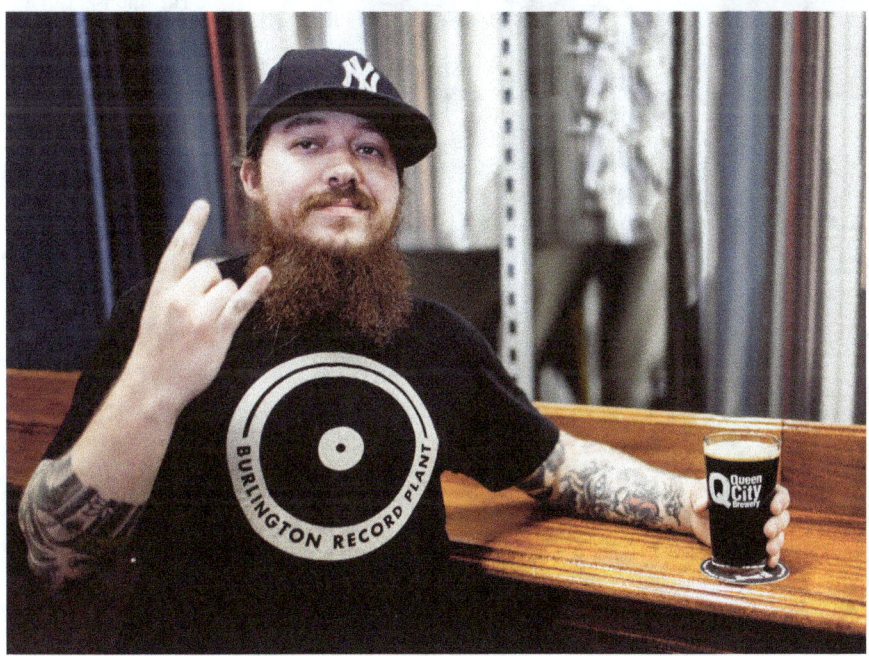

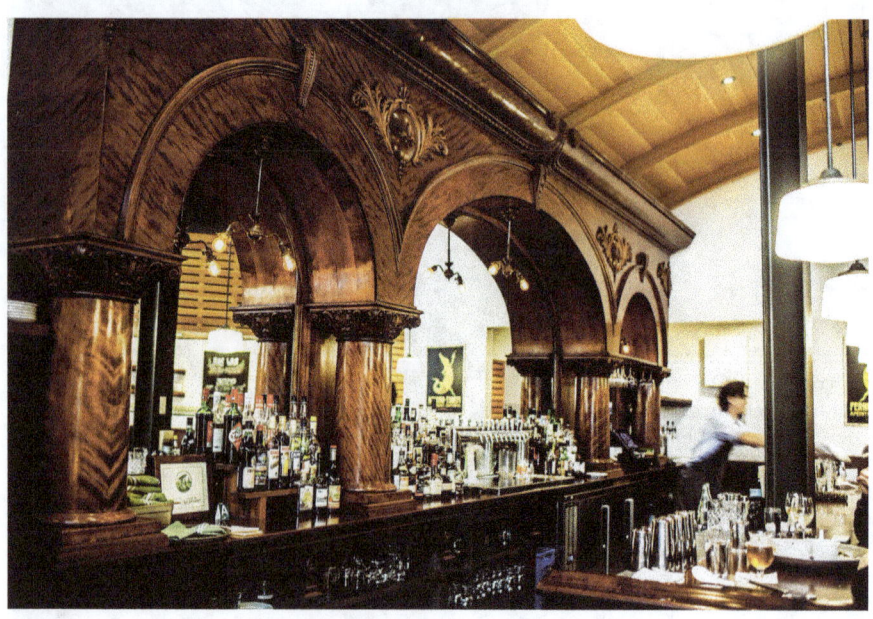

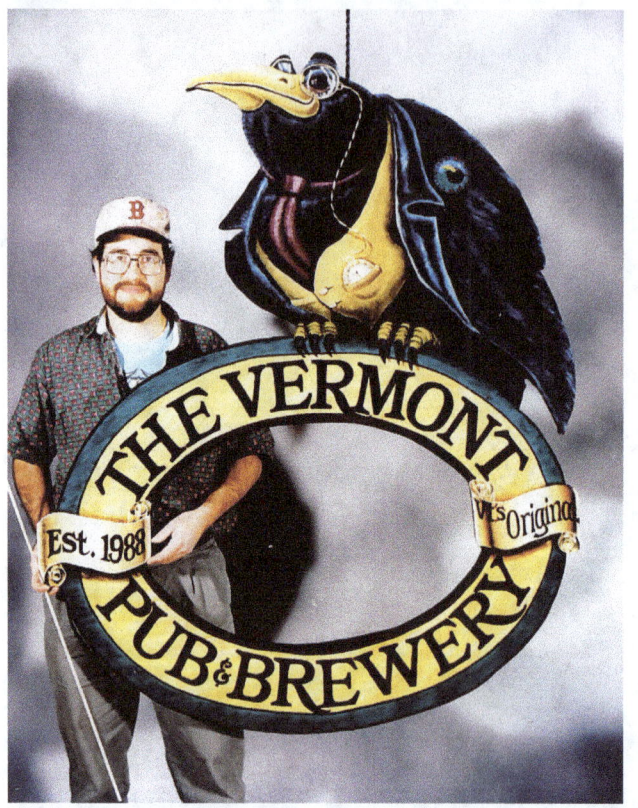

Above: The bar at the Great Northern on Pine Street has a long history. On the right, barman Colin Walsh, formerly of the Farmhouse Tap & Grill, is serving beers. *Courtesy Paul Sarne.*

Left: The Vermont Pub & Brewery sign carved by Jim DiStefano. *Courtesy Jim DiStefano.*

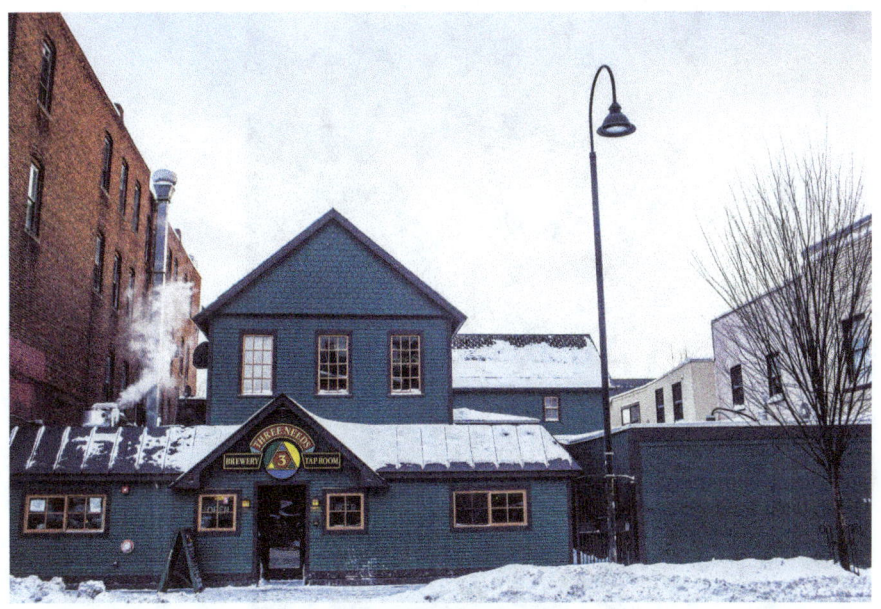

Three Needs Taproom on Pearl Street. *Courtesy Paul Sarne.*

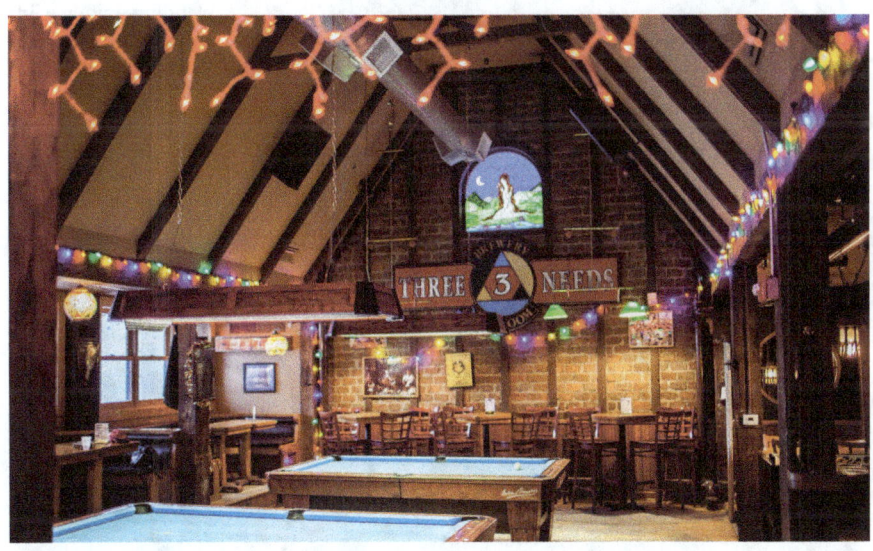

The custom stained-glass window above the pool hall at Three Needs was installed by the previous tenants. *Courtesy Paul Sarne.*

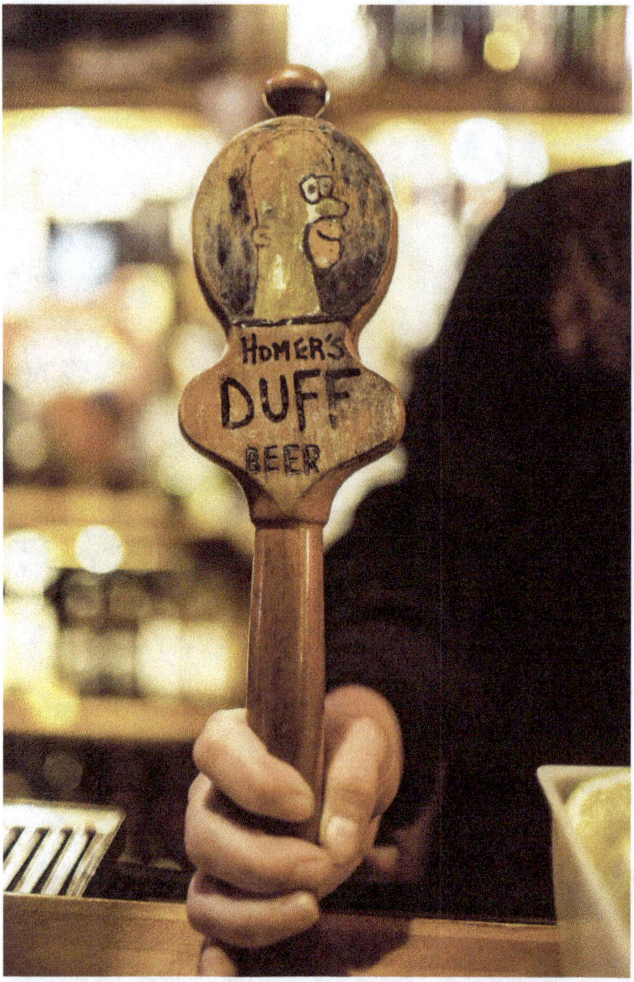

Left: The original hand-painted "Duff" tap handle. Three Needs offers a special keg each day for one dollar per pint until the keg is gone, calling the beer "Duff"—a reference to *The Simpsons*. *Courtesy Paul Sarne.*

Below: House of Fermentology specializes in blending unique wild and sour ales that have been aged in oak barrels. *Courtesy Bill Mares.*

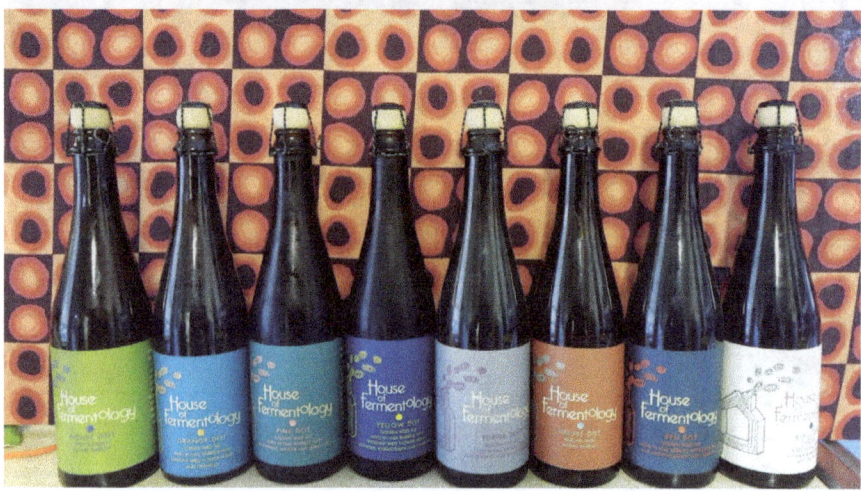

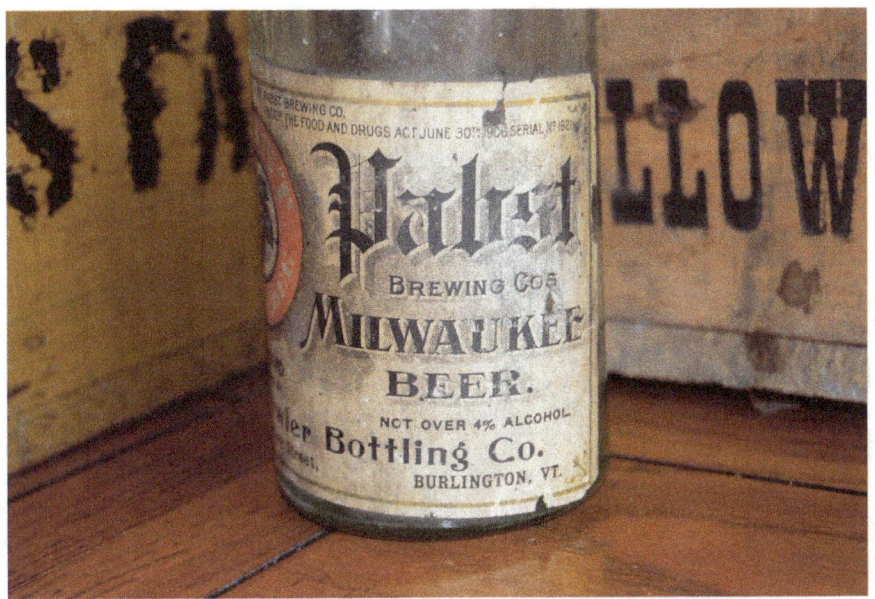

A Prohibition-era bottle of Pabst beer bottled in Burlington, Vermont. Note the ABV and additional language on the label. *Courtesy Kurt Staudter.*

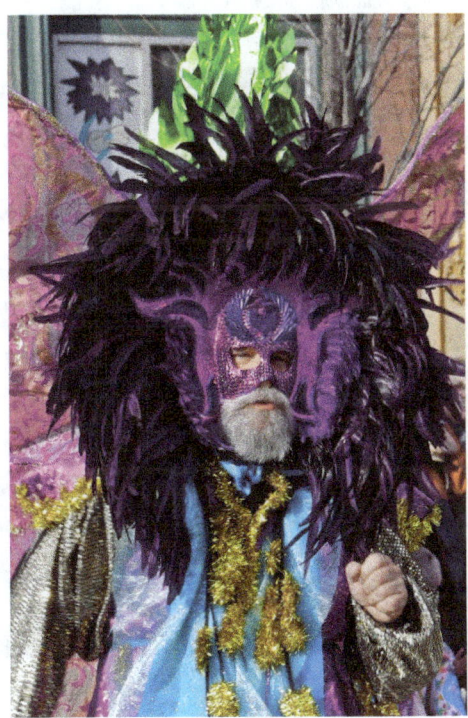

Alan Newman, cofounder of Magic Hat Brewery, dressed up to lead the 2010 Mardi Gras parade. *Courtesy Paul Kowalski.*

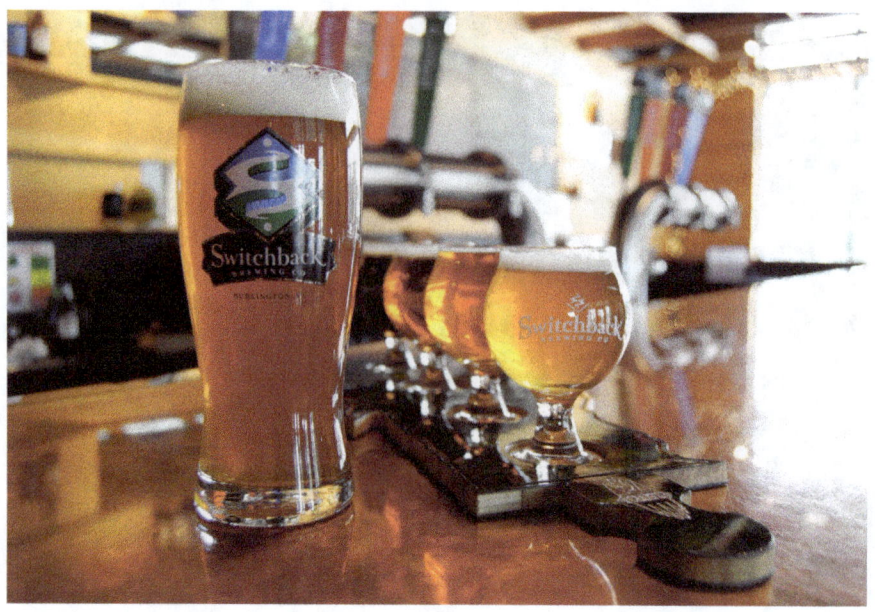

The Switchback Brewery taproom has become the anchor of "Pint Street," despite being farther west on Flynn Avenue. Visitors can enjoy samples, pints and McKenzie hot dogs. *Courtesy Switchback Brewing Co.*

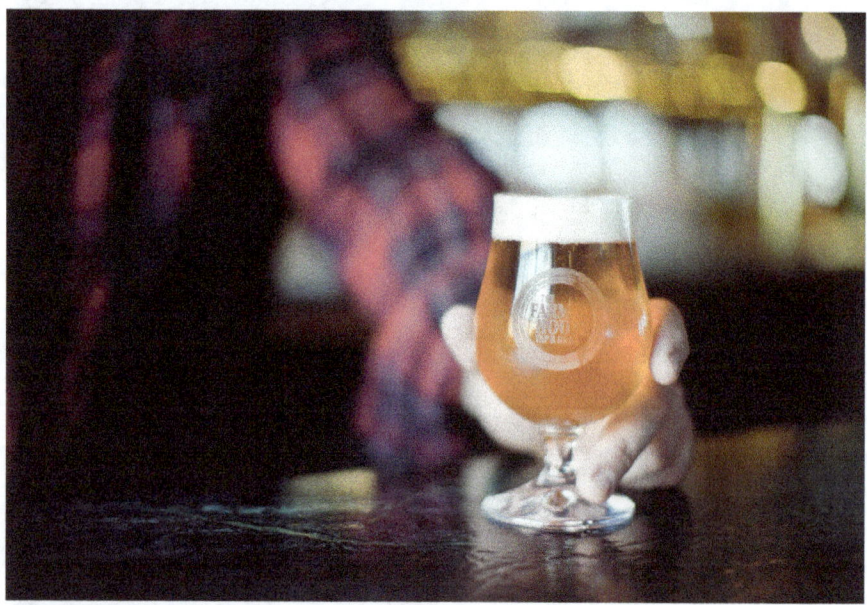

The Farmhouse Tap & Grill espouses the concept that good beer service requires using the proper glassware for each style. *Courtesy Paul Sarne.*

Ask any brewer, and they'll tell you that cleaning is the most important part of the brewing process. *Courtesy Switchback Brewing Co.*

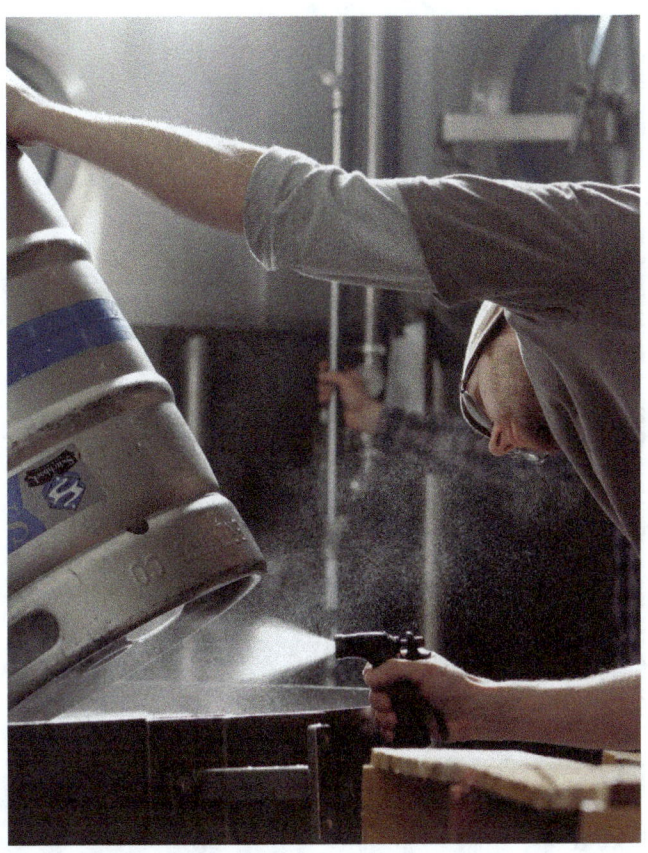

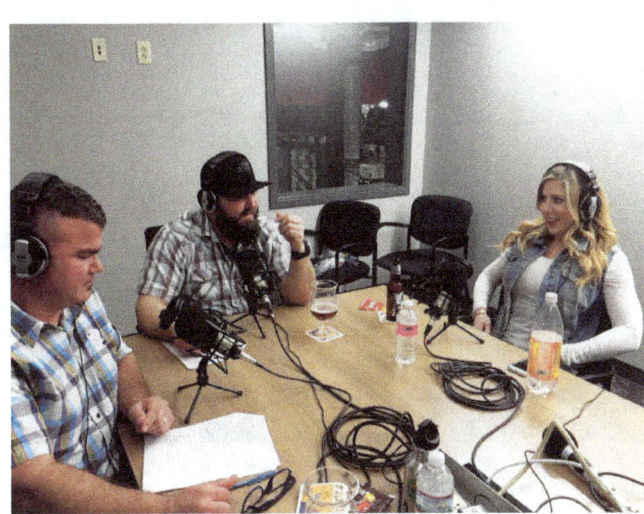

Jason Strempek (*left*) and Jeff S. Baker II (*center*) interviewed Brittney Hibbs-Kelson (*right*) for an episode of the *It's The Beer Talking* podcast. *Courtesy Ryan Chaffin.*

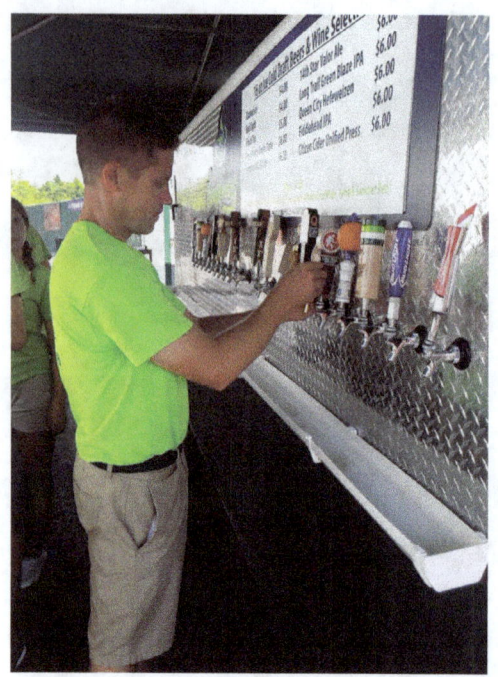

Burlington is serious about craft beer—so much so that the Vermont Lake Monsters minor-league baseball team offers local crafts on draft. *Courtesy Farrell Distributing Corporation.*

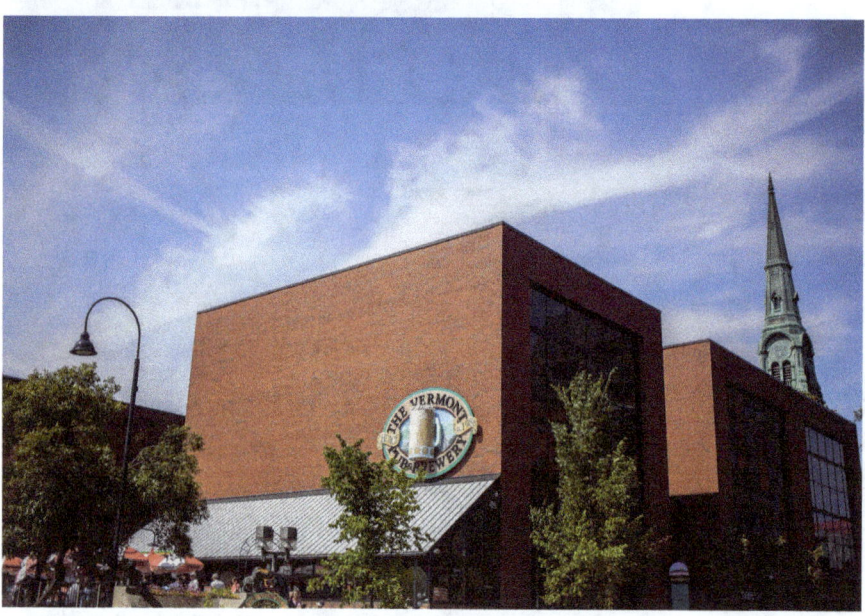

The facade of Vermont Pub & Brewery in 2018. *Courtesy Paul Sarne.*

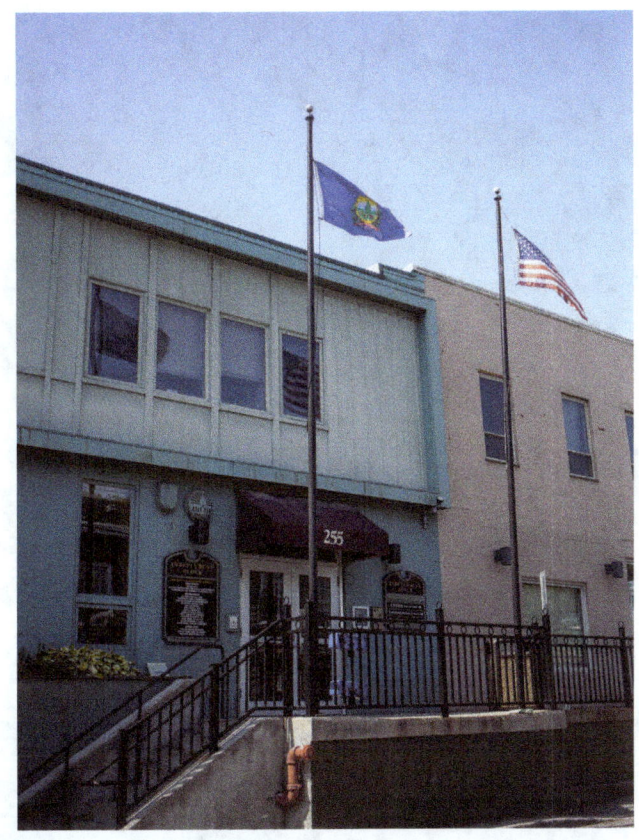

Right: This is the site of Burlington Brewing Company, which closed around 1885. The building has been demolished and reconstructed. *Courtesy Paul Sarne.*

Below: Zero Gravity Craft Brewery commissioned this mural of a hummingbird from artist Mary Lacy. A hummingbird is featured on the Zero Gravity logo. *Courtesy Paul Sarne.*

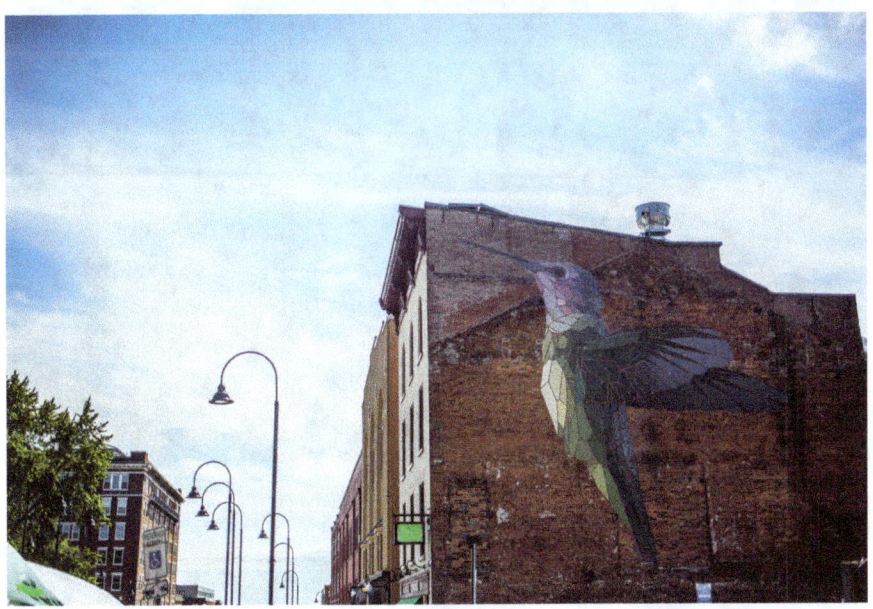

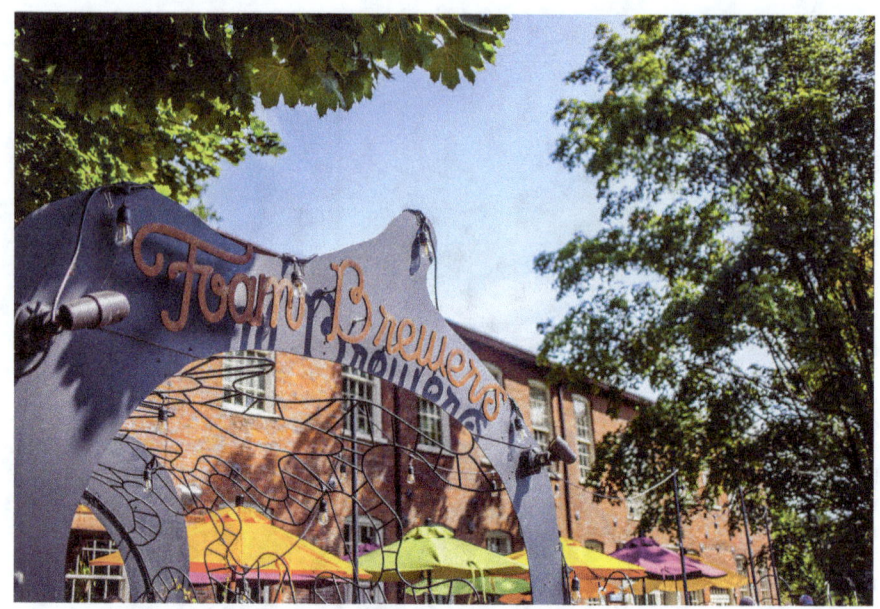

The entrance to Foam Brewers on Burlington's waterfront. *Courtesy Paul Sarne.*

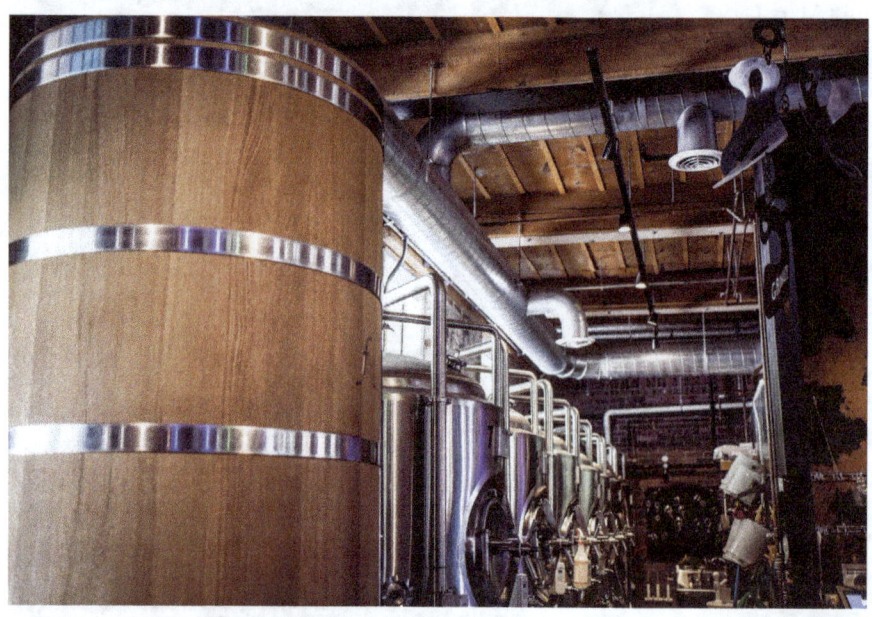

Fermentation tanks at Foam Brewers, including a wooden foeder (*left*) for fermenting wild ales. *Courtesy Paul Sarne.*

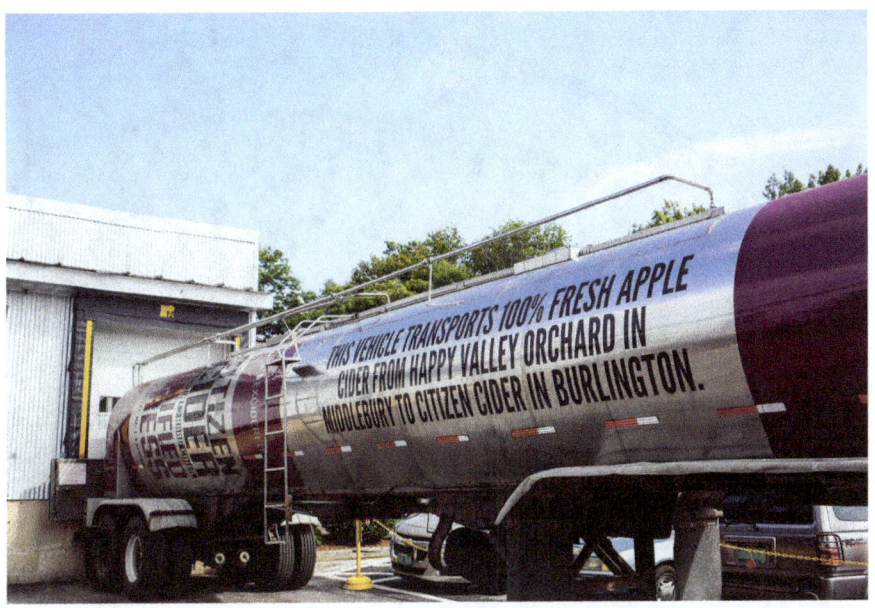

Citizen Cider, on "Pint Street," commissioned their tanker truck to look like a giant version of their cans. *Courtesy Paul Sarne.*

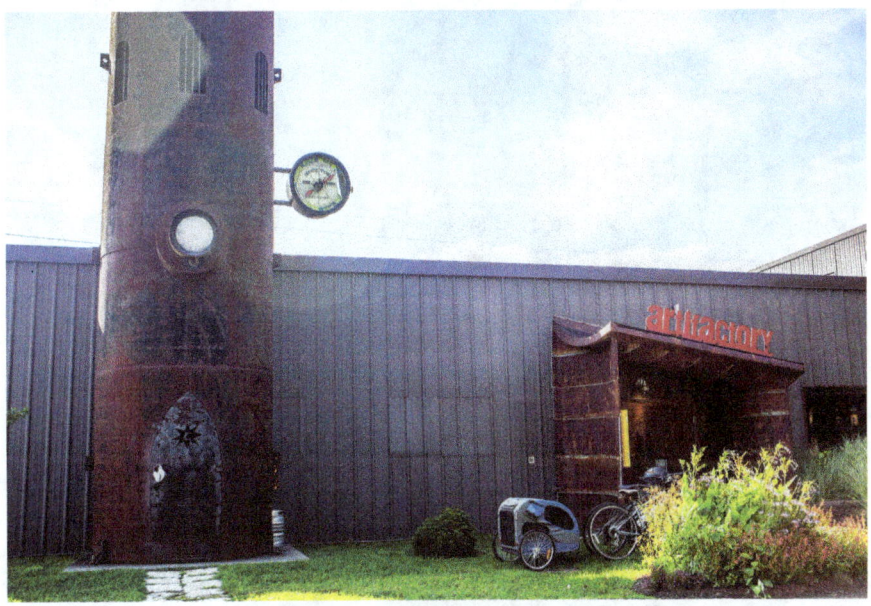

The entrance to Magic Hat's Artifactory was designed by Russ Bennet, founder and owner of NorthLand Visual Design. *Courtesy Paul Sarne.*

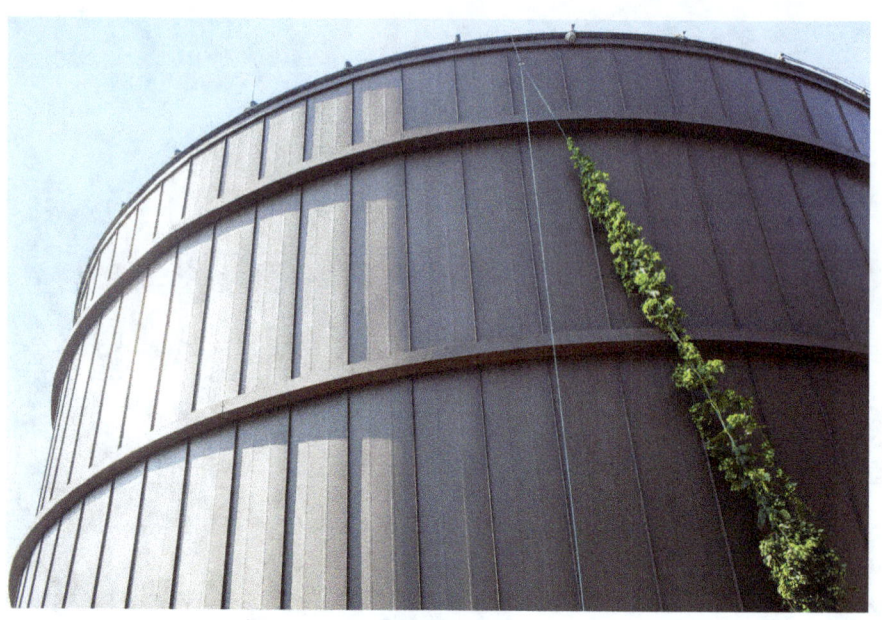

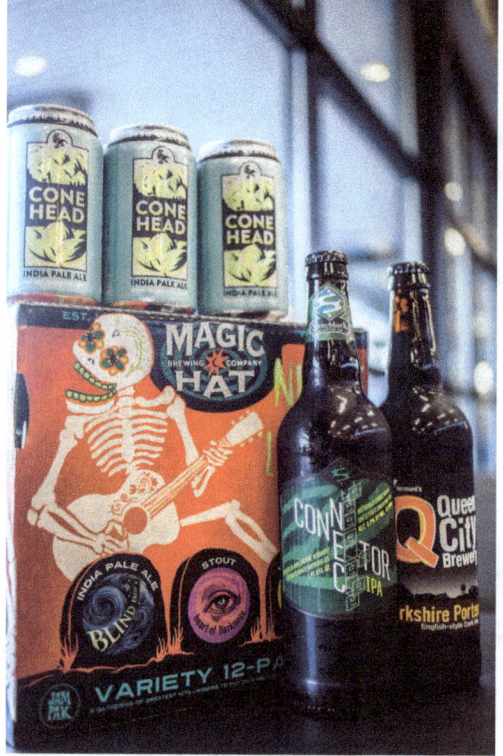

Above: A hop vine growing up the side of Magic Hat's Anaerobic Digester. *Courtesy Paul Sarne.*

Left: Beer comes in all shapes and sizes in Burlington. *Courtesy Paul Sarne.*

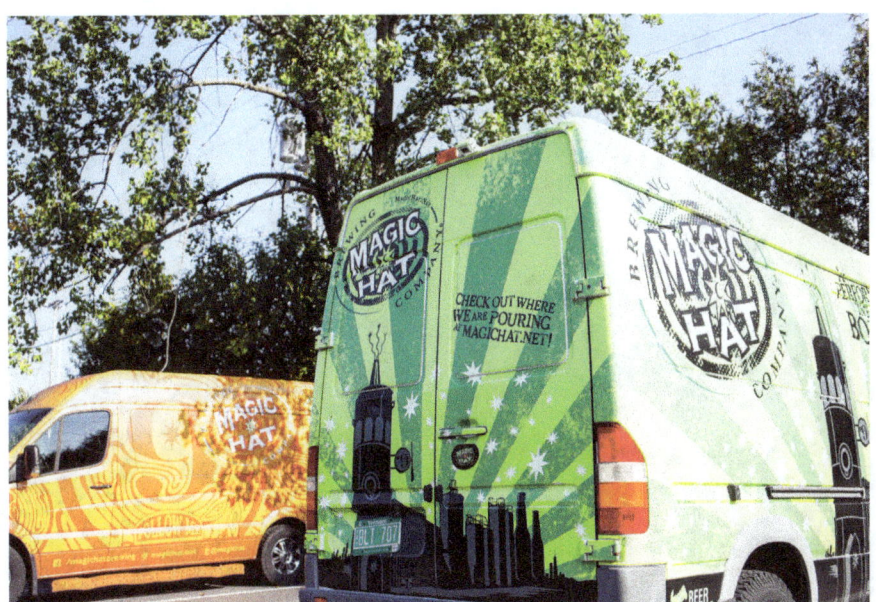

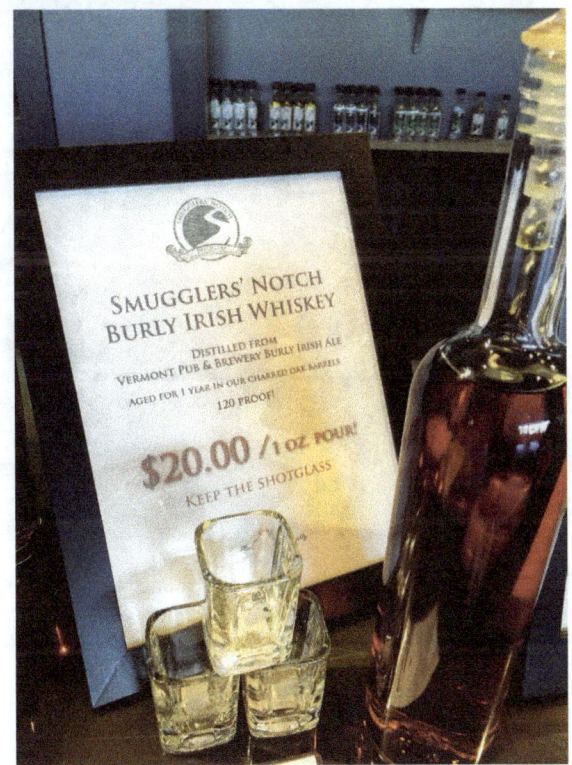

Above: Magic Hat Brewing Company is known for their visually stimulating artwork and designs. *Courtesy Paul Sarne.*

Right: Vermont distilleries and breweries have collaborated on projects, including this whiskey distilled from Vermont Pub & Brewery's Burly Irish Ale. *Courtesy Jeff S. Baker II.*

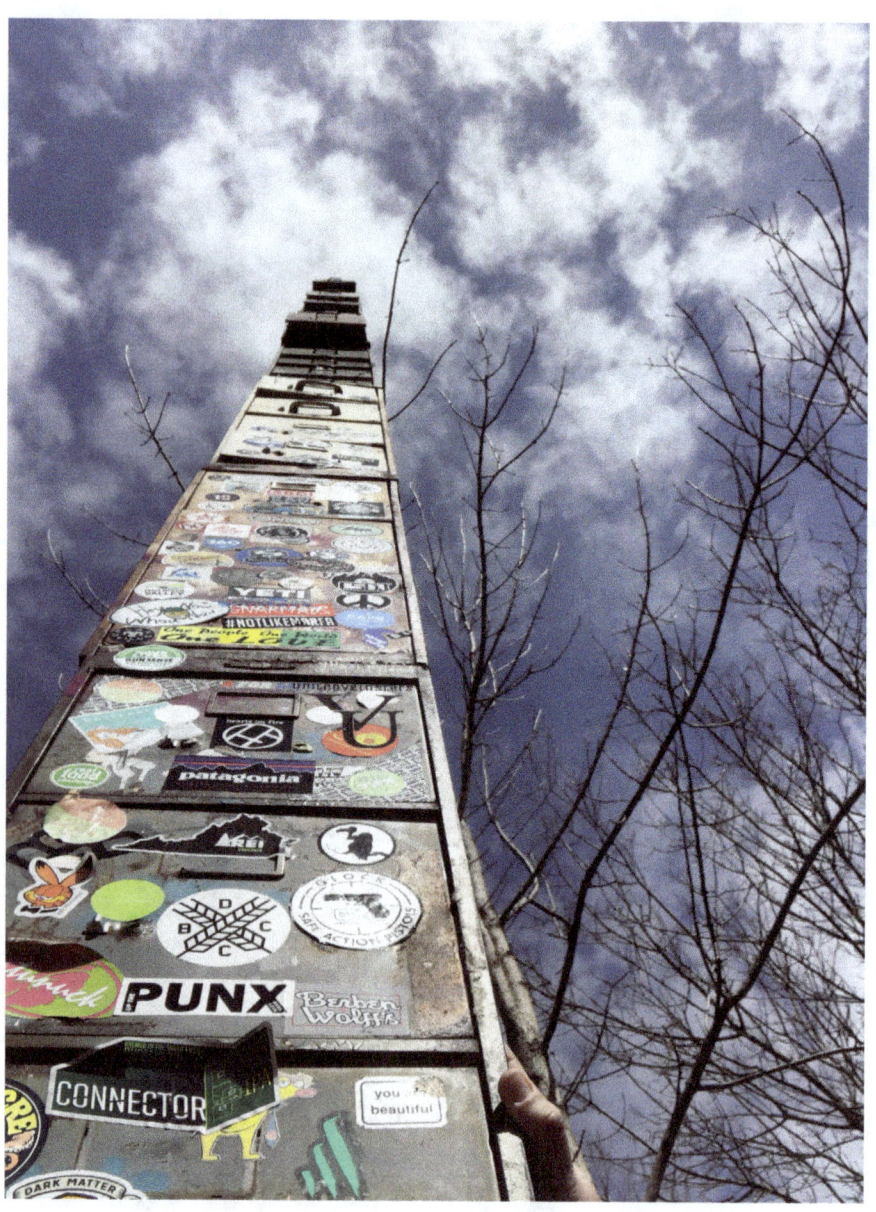

Burlington is home to "the world's tallest filing cabinet." This art installation on Flynn Avenue was created by architect Bren Alvarez and inspired the design of the tap handle for the Switchback Connector IPA. *Courtesy Jeff S. Baker II.*

SWITCHBACK BREWING COMPANY

160 FLYNN AVENUE, BURLINGTON

*"I tasted the beer and thought…this beer is good enough
to make a living off of."*[141]
—*Bill Cherry, cofounder and brewmaster*

Switchback Brewing Company was founded in 2002 on Flynn Avenue in Burlington, next door to McKenzie Country Meats, a century-old producer of cold cuts, cured meats and hot dogs. Despite setting up shop in this rather out-of-the-way industrial space, Switchback has firmly established itself in Vermont's—and, for that matter, New England's—brewing scene. For a long time, Switchback was an enigma for craft beer drinkers, since **Switchback Ale**, their flagship beer, was a draft-only offering, and it wasn't easily categorized. But many craft beer fans have spoken fondly of tracking down their first pint of Switchback Ale in the Green Mountain State.

EARNING HIS LETTERS

Brewmaster and owner Bill Cherry and his business partner Jeff Neiblum moved on the idea of opening a brewery in 2002 with equipment setup and test batches taking place through August and September. The first pint of Switchback Ale was served at Ake's Place, a sports bar and longtime staple

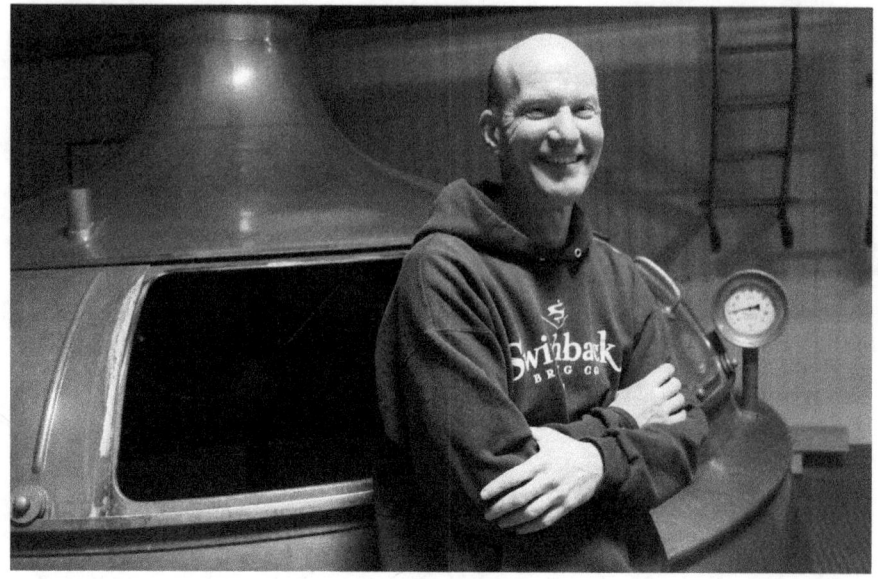

Bill Cherry, brewmaster of Switchback Brewing Company, standing in front of their German copper brewhouse. *Courtesy Switchback Brewing Co.*

on Church Street in Burlington, on October 22, 2002. Achieving that pint was no easy task for Cherry. It took many years and twists in the road to get to that point.

Cherry desired to join the brewing world but struggled to break into the business. Attempting to gain insight and set a plan to break into the industry, he sent letters to executives in large brewing companies about potential open positions and what he would need to get hired. Figuring that brewers and plant managers were tied up and busy juggling the operation, Cherry guessed that the higher-ups were more likely to have time to respond. An executive from Coors went as far as to respond with a personal letter and resources from the Coors brewing library, plus a list of brewing schools. Bottom line? Cherry needed more education. That became the blueprint.

Cherry's next stop on the road to Switchback was brewing school at University of California, Davis, where he earned a master's degree in brewing. He figured he was going to end up in a large-scale production brewery, and he was even told by his professor that he was going to make a great corporate brewer. But many of his classmates were discussing brewpubs, a relatively new phenomenon. Although Cherry felt comfortable in the environment of packaging, quality control and brewing, thankfully for the Vermont beer scene, he was never sucked in by the big boys of the brewing industry.

QUALITY OVER QUANTITY

Upon landing at Boulevard Brewing in Kansas City, Missouri, in the mid-1990s, Cherry's job was to "translate the brewer's art into a commercial product." Boulevard's owners were dedicated to producing quality beer. Cherry was quick to note the role quality played in the late '90s, as many breweries at the time focused on expansion, taking their eyes off some of the quality standards needed to be successful. Needless to say, brewery closures occurred in rapid succession. The iconic Vermont brewery Catamount, the first brewery in Vermont since federal prohibition, was eventually a casualty of this failing. However, brewers who focused on production and quality survived this period, often thriving even while others were closing.

The Switchback plan was to let growth happen organically rather than attempt to navigate a five-year strategic growth plan. In the beginning, Cherry handled all aspects of the business, but he started to become overextended, spending seventeen hours a day at the brewery. Toward the end of 2003, he hired the brewery's first employee: Chris Dooley.

DULY NEEDED

The hiring of Dooley allowed Cherry to focus a bit more on marketing and selling Switchback Ale. Although it was still only available on draft, it started to garner high demand, pushing to the point that even with more employees, the brewery hit capacity and required twenty-four-hour operations just to keep up. In 2005, Cherry hired Gretchen Langfeldt, who "called once a month for five months until [she] annoyed Bill enough" to let her wash kegs once a week.[142] With a degree in mechanical engineering, Langfeldt would eventually become plant engineer. A year later, Cherry hired Tony Morse, who has put his degree in education to use at the brewery, helping the team to learn and grow together.[143]

Staying true to his original plan, now that the market demand warranted expansion, Cherry started building out a new brewhouse in 2008 and completed it in 2010. A 1964 vintage copper brewhouse from Brauerei Schmucker of Beerfelden, Germany, was acquired, as well as new tanks and fermenters, vastly expanding Switchback's capacity. Todd Haire, a former brewer at Switchback and now the owner of two Burlington breweries,

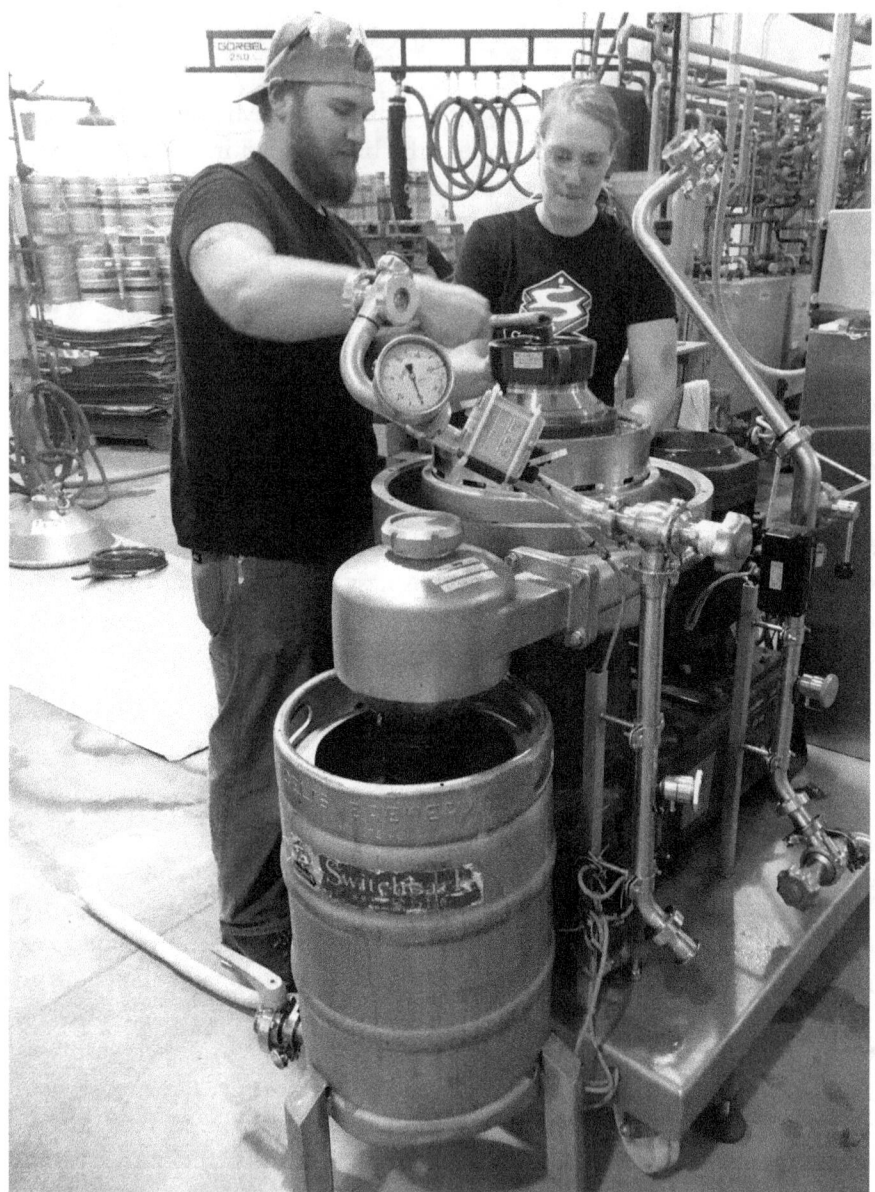

Switchback brewer Cody Smith (*left*) and plant engineer Gretchen Langfeldt (*right*) making equipment adjustments. *Courtesy Switchback Brewing Co.*

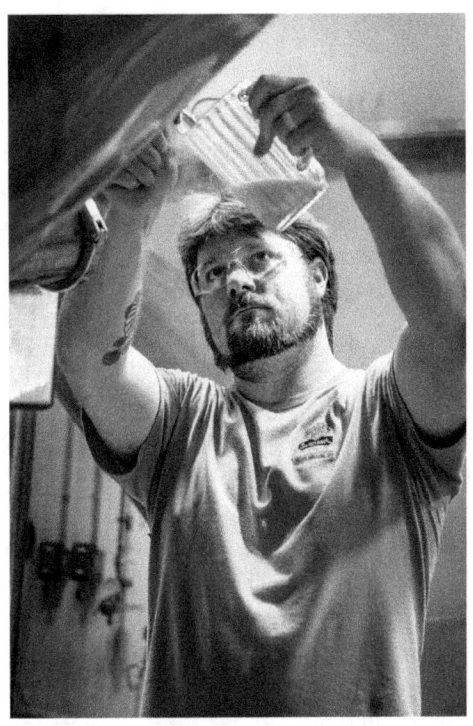

Head brewer Tony Morse pulling a beer sample from a tank at Switchback Brewing Company. *Courtesy Switchback Brewing Co.*

spoke highly of Cherry's approach, saying Cherry is "very methodical about putting in equipment that's needed to do what needs to get done."[144]

The completed expansion led to something surprising: new recipes in bigger batches. In quick succession, Switchback's offerings started to increase, with the brewery presenting new ales: **Roasted Red, Slow-Fermented Brown Ale, Dooley's Belated Porter** and **Extra Pale Ale**. As sales grew, Cherry decided it was time to acquire a bottling line, sourcing a top-of-the-line German-engineered filling system that is considered one of the finest systems available: "I don't understand brewers' blind spot in packaging. They worked so hard to make such a quality product, and it does not hold up from their packaging."[145] Quality brewers want to ensure that consumers get to taste the *real beer* the way the brewer intended. This was a driving factor in Switchback Brewing's long delay in bottling its beers. Until the packaging system and quality met Cherry's standards, it was simply not going to happen. Thankfully, it did, with the first releases occurring in twenty-two-ounce "bomber" bottles, then, later, in six-packs of twelve-ounce "stubby" bottles. In 2018, the brewery even added four-packs of sixteen-ounce cans—a popular national trend.

SUI GENERIS

Switchback Brewing Company is unique in the beer scene, offering many styles that other breweries do not. Most of the beers produced by Switchback do not adhere to traditional style guidelines, adding to its sui generis status. The flagship brew, Switchback Ale, is a combination of all the attributes that brewmaster Cherry liked in other beers. It's somewhat amorphous; it's unfiltered, 100 percent naturally carbonated and sits somewhere on the spectrum between a pale ale and an amber ale. The philosophy of Switchback has always been to not follow what the industry was doing but rather to brew what Cherry wanted to drink. That philosophy has stayed the same, with Cherry noting, "we are trying to add to the brewing culture, not duplicate it." The market has clearly shown that Cherry isn't the only one who desired this sort of beer.

Then came the addition of **Connector IPA** to the lineup. After Switchback had long avoided the style, which accounts for nearly 25 percent of all craft beer sales (according to the Brewers Association), the creation of this beer stemmed from demand in the Switchback taproom. Cherry noted that from a business perspective, his customers wanted an IPA, so he should fill that void. He sent a brewer incognito to other Vermont breweries, including Fiddlehead Brewing in Shelburne, Vermont, to bring back research samples.[146] In a way, the creation of Connector IPA was another extension of Cherry's business philosophy—when sales and demand are there, then it's time to expand.

SELLING THE COMPANY

Perhaps one of Cherry's biggest contributions to the Vermont beer scene came in February 2017. Cherry announced that he was selling the company—to the employees. Switchback Brewing Company became Vermont's first ESOP (employee stock ownership plan) brewery. The move made Switchback entirely employee-owned, meaning the brewery would stay firmly planted in Vermont. "I wanted Switchback to be Vermont-owned forever," Cherry wrote in a press release. "Entrusting the brewery to the employees ensures the company will carry on with its mission to provide great beer and great jobs to the local community."

Switchback Brewing Company epitomizes the ethos of the Vermont brewing scene. The company grew through grassroots efforts, they are dedicated to the local community and they have never once wavered in their commitment to quality. Although their beers are difficult to categorize stylistically, one thing is certain: they have earned a large fan base throughout Vermont and New England that shows no signs of abating.

AMERICAN FLATBREAD BURLINGTON HEARTH

115 SAINT PAUL STREET, BURLINGTON

"A gathering place for our community to celebrate the gifts of local food and craft beer."[147]

When Rob Downey was in law school and Paul Sayler was pursuing a new career in brewing, the two teamed up to start a brewpub and started scouting locations, setting their sights on Burlington, Vermont. After signing a deal for a space in a new building on Lake Street that would eventually be dubbed Main Street Landing, area residents sued the building project and held it up in court for two years. It would be a while before Downey and Sayler would have their brewpub up and running, but the future turned out to be very bright.

THE BREWER'S BEGINNING

Paul Sayler has been a pillar in the Vermont beer community since the 1990s. Armed with a bachelor of arts degree in philosophy from Hampshire College and experiences as varied as working for a railroad company near Seattle, as well as having done a bike tour of German breweries, Sayler came to Vermont to intern at Catamount Brewing Company.

His three-month internship was mainly washing kegs, cleaning tanks and assisting on the bottling line. But he showed potential, and at the end of his

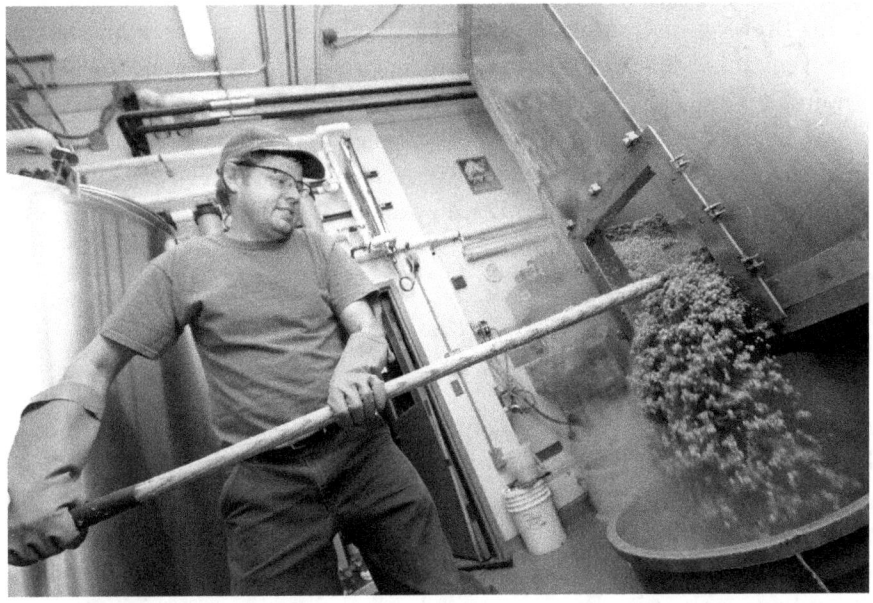

Brewmaster Paul Sayler brought taproom culture to Burlington by cofounding American Flatbread Burlington Hearth, Zero Gravity Craft Brewery and the Farmhouse Tap & Grill. *Courtesy Zero Gravity Craft Brewery.*

internship, he was offered a full-time job. He showed so much potential, in fact, that his employers offered him a scholarship to attend the Siebel Institute brewing program, but Sayler declined out of respect, as he didn't think he'd be with the brewery long enough for it to be worth their investment.

Six years later, Sayler had earned an on-the-job degree at Catamount in fermentation, quality control and virtually all aspects of brewing. Eventually, he decided it was time to spread his wings a bit, so he left to foster his professional growth.

After brewing at Colorado Brewing Company in Danbury, Connecticut, Sayler did a stint at Commonwealth Brewing Company in Boston and then at their Rockefeller Center location in New York City. As luck would have it, it was at the Rockefeller Center location that Sayler met a tall, red-haired Texan named Todd Haire, who also eventually ended up brewing in Burlington.

Between Sayler's time at Catamount, where he learned of the importance of having an on-site laboratory, and brewing in New York City, Sayler fell in love with the "canon" of European brewing tradition. Inspired by Garrett Oliver (Brooklyn Brewery) and Phil Markowski (Two

Roads Brewing in Connecticut), Sayler credited both as "very important people in [his] development because they're both, stylistically speaking, erudite individuals."[148]

But, before Sayler could open his own brewpub, there was lots of work to be done, including brewing at the Bobcat Café in Bristol, Vermont, a project done in collaboration with Rob Downey and Robert Fuller.

EARLY DAYS

After calling the owners of American Flatbread in Waitsfield, Vermont, Downey and Sayler worked out a plan to collaborate with them on their new project. Sayler spent time learning the flatbread trade in Waitsfield, then helped open another location in Middlebury, as well as a production kitchen for their retail flatbreads. While the litigation held up the Main Street Landing location, the duo started to look for other potential locations and settled upon a three-story building constructed around 1884[149] at 115 Saint Paul Street, the former home of Carbur's Restaurant.

AN ASIDE: CARBUR'S RESTAURANT

Founded in 1974 by Carl L. Capra and Burr Vail, Carbur's Restaurant became known for its extensive food menu, which sometimes reached a length of sixteen pages,[150] and their extensive beer list. Paul Limberty, owner of Dragonfly Sugarworks, recalled that Carbur's had a "beers around the world card." Limberty—and other beer fans—were exposed to imported beers from places like Denmark, Germany, China, Japan and Belgium. Those who completed the card received a free gift from the restaurant.[151]

Carbur's Restaurant eventually expanded to other locations, including in Plattsburgh, New York; Portland, Maine; and Hadley, Massachusetts. In early 1984, Carbur's purchased the Déjà Vu Café on Pearl Street in Burlington, which eventually became the home of Three Needs Taproom & Brewery. In 1985, Capra and Vail had a falling out and dissolved their partnership. On October 4, 1985, Capra was found dead in his home from an apparent self-inflicted gunshot wound.[152] In 2002, the Burlington Carbur's closed its doors.

Capra's death is often conflated with another tragedy. On January 9, 1992, shortly after arriving for work, a twenty-five-year-old cook named David J. Mario went down into the basement and shot himself in the head.[153]

HAUNTED

There have been numerous reports of paranormal activity at 115 Saint Paul Street, particularly in the basement. A handful of American Flatbread Burlington Hearth's female employees have reported feeling like they were being pinched or groped in the basement when no one else was around. At least one employee reported that after entering a walk-in cooler, the door slammed shut, and loud banging hammered the cooler from every side. The door eventually gave way, and when the employee rushed back upstairs, no one else had heard the commotion. Another incident occurred at the bar: a bartender had just finished washing the last glasses for the night, and when this person turned their back to the glasses, they heard a sound like a wind chime. Turning back around, the bartender saw that the clean glasses were stacked in a neat pyramid on the bar, and no one else was in the room.[154] Whether 115 St. Paul Street is haunted by the ghost of Carl Capra, David Mario or perhaps by a cadre of pirates from Lake Champlain,[155] these authors cannot say for sure.

TAPROOM CULTURE

The concept behind American Flatbread Burlington Hearth was to fuse an already successful restaurant concept with a taproom. Sayler identified that the beer scene in Burlington was missing a true taproom. Of course, Greg and Nancy Noonan brought the first brewpub to the Queen City, and Sayler often visited it while he was brewing for Catamount. But to Sayler, there was a difference in what he wanted to do for his own project.

American Flatbread Burlington Hearth would become "a taproom with house beers."

"That may sound like semantics," Sayler said in an interview, "but it's alternative is 'brewpub with guest beers'…a business where the owner has forced the brewer to have other beers and the brewer hates it."[156] (For more, see the chapter on taprooms.)

To complement the house-brewed beers, Sayler reached out to importers like the Shelton Brothers, an importer of small, independent and often esoteric European beers, and Merchant du Vin, an importer of traditional British and Belgian beers. He worked with them to bring Europe's finest beers to Vermont so he could put them on tap right next to his house-brewed beers. The taproom is about educating consumers and stretching their palates. The best way to do that, Sayler figured, was to put world-class beers right in front of fledgling craft beer drinkers.

When the popularity of the house brand, Zero Gravity, grew to a point of needing more draft lines, Sayler and Downey would transfer the taproom concept to a new business venture: The Farmhouse Tap & Grill.

SETTING THE BAR

The buildout for American Flatbread Burlington Hearth resulted in some impressive internal features. In addition to the handmade wood-fired oven made from local red rock, the bar itself is a thing of beauty. Working with a Philadelphia broker, Sayler and Downey bought a vintage elaborate backbar from the Midwest. With large mirrors, carved columns, Edison-style lightbulbs and elegant vintage opaque green-and-white glass inserts along the top, it's a real treasure to behold.

Jeff Baumann, the bar manager at American Flatbread Burlington Hearth, employed his calm demeanor and unpretentious approach while sharing beer knowledge with patrons, which afforded him a long career with the Zero Gravity group. In 2010, Baumann moved to Portland, Oregon, to scout a new location for American Flatbread and Zero Gravity, but the plan never came to fruition. When his friend and former bartender, Chad Rich, then the bar manager at the Farmhouse Tap & Grill, decided to open a brewpub in Waterbury, Baumann came on as bar manager of Prohibition Pig. But when Zero Gravity opened its Pine Street location, Baumann came back into the fold and was instrumental in opening later projects, including Monarch & the Milkweed and the Great Northern.

Today, Ryan McFarlin is the bar manager at American Flatbread Burlington Hearth.

First Beers

In a nod to his love for European beer styles, Sayler launched his in-house Zero Gravity Craft Brewery menu with beers like a Munich-style helles, an extra special bitter, an extra stout, a porter and **TLA** IPA. TLA became an ever-evolving IPA recipe that Sayler would tweak and adjust from batch to batch, saying that it was "always in flux." Heraclitus would be proud.

Destiny Arrives

In 2008, around the time they were gearing up to open Farmhouse, Sayler hired a young brewer with a degree in geology to help develop new recipes. Her name was Destiny Saxon, and she would flourish and grow with the brewery, taking it to the next level.

Saxon got her start like many professional brewers: by making beer at home as a hobby and not loving her day job. She left a career in an environmental laboratory and landed a job brewing for Otter Creek Brewing Company and Wolaver's Organic in Middlebury, Vermont. During her four years at

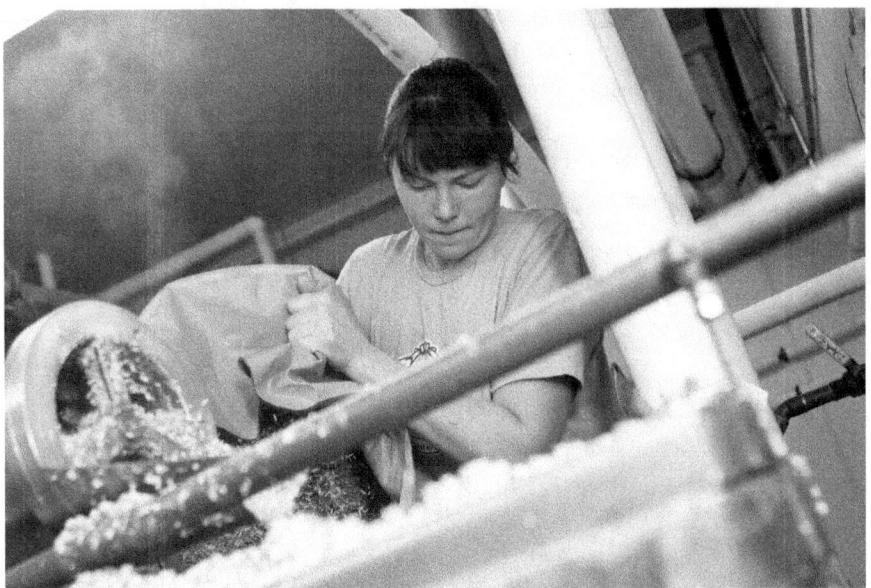

Destiny Saxon, brewer at Zero Gravity Craft Brewery, mashing in. *Courtesy Zero Gravity Craft Brewery.*

Otter Creek and Wolaver's, Saxon attended the Siebel Institute. One day, she met Sayler, and when he was in need of a brewer, Saxon convinced him to give her a full-time gig at Zero Gravity.[157]

Because Saxon demonstrated the same brewing philosophy as Sayler, he stepped aside and named her head brewer. Sayler, who would retain the title of brewmaster, wanted to give Saxon creative freedom and space to "own" her beer recipes and experiment as she saw fit.[158]

GRUITS AND DRUIDS

Zero Gravity Craft Brewery at American Flatbread Burlington Hearth has earned a name for itself in the Vermont beer community for various reasons, not least of which is their celebration of gruit ales. Gruits, based on a historical European style of beer brewed with herbs instead of hops, were not only heralded for their taste but also for their healing or enlivening properties.

It is perhaps only fitting that Zero Gravity produces these pseudo-medicinal beers at 115 St. Paul Street, as the building was home to a patent medicine company owned by Nathaniel King Brown in the 1880s.[159]

For the winter and summer solstices, Zero Gravity produces unique gruits using herbs from Burlington's Hallow Herb Farm, including yarrow, sweet gale, woodruff, Labrador tea, mugwort and nettles.[160] In some years, Zero Gravity held a celebration they called the Running of the Druids. Patrons and employees would dress in brown robes and parade a cask of gruit from the top of Church Street through downtown and right into Burlington Hearth, where the cask was ceremoniously tapped.

ACCOLADES

In addition to lots of local praise, Burlington Hearth has earned recognition from national press such as the *New York Times* and *Vogue*,[161] as well as international publications. RateBeer.com named American Flatbread Burlington Hearth as the best brewpub in Vermont five times between 2012 and 2018. Zero Gravity Craft Brewery was nominated as a finalist in the 2018 *Seven Days* Daysies Awards for both "Best Craft Brewery" and "Best Beer from a Local Brewery" (for **Conehead IPA**).[162]

PRODUCTION BREWERY

When Zero Gravity maxed out its brewing capacity, there was still a growing demand at other Vermont watering holes for their beer. The decision was made to build an off-site second brewery that would handle all of the production brewing, while the small-batch, more experimental brewing would still happen at Burlington Hearth. The new production brewery was built at 716 Pine Street in Burlington's South End and opened in 2015.

ZERO GRAVITY CRAFT BREWERY

716 PINE STREET, BURLINGTON

A s the accolades continued to roll in and their popularity grew, it became difficult for Paul Sayler and Destiny Saxon to keep up with the demand for Zero Gravity beers. A second brewery was needed, but bringing this to fruition was not as easy as flipping a switch.

PRE–PINE STREET

Before they could roll out a new brewery, they needed a stopgap solution. Sayler rang his friends at Smuttynose Brewing Company in New Hampshire, and they agreed to brew and keg **Conehead IPA** and **Green State Lager** for the Vermont market. Conehead became the most popular offering from Zero Gravity, with the light blue cans available at almost every beer shop, convenience store and grocery store in the Green Mountain State by 2018. Green State Lager, a pilsner somewhere between the Bohemian and German styles, has chased Conehead's success.

NEW DIGS

In the spring of 2015, the new production brewery on Pine Street was ready to open to the public. Armed with a brand-new thirty-barrel brewhouse, tons of warehouse space for expansion and their own canning line, Zero Gravity's brewing team was ready to rock and roll.

Although the labyrinth of shiny stainless steel tanks is a seriously impressive vision to behold, the tasting room, which faces Queen City Brewery across the street, offers a master class in modern design. White subway tiles line the backbar, which is only interrupted by a line of tap faucets, and right at dead center, the head of a firkin cask quixotically juts out, allowing for the proper serving of gravity casks.

On one wall is a massive thirty-two-foot-long mural by Blaine Fontana, an artist from Portland, Oregon. It depicts a black-capped chickadee, which is native to Vermont, roosting on a multicolored Seussian tree branch.[163] On the other side of the tap room is a retail space offering cold beer to go and branded merchandise. Just beyond the retail area is a cozy, lounge-like room dotted with comfortable seating, a small wood stove, Persian carpets and a full-sized shuffleboard table. It would be easy to forget that you're in a brewery when lounging in this space—it feels more like a friend's den.

PRODUCTION BEERS

With this new full-time, large-capacity brewing facility, the Zero Gravity team has produced dozens of beers that have been distributed throughout Vermont and New England. While Conehead IPA and Green State Lager are the two top sellers, other Zero Gravity releases have made waves in the beer scene.

Little Wolf, an American pale ale, gets its name from the Latin term for hops: *Humulus lupulus*. Zero Gravity brewers used an enzyme in the brewing process to reduce and denature gluten protein chains in the final beer, rendering it below the FDA threshold of twenty parts per million so it can be labeled "gluten-free."

Zero Gravity has released a rotating series of double IPAs in cans, including **Madonna, Grand Royal** and **Choice Make Good**.

In a nod to the 2004 Wes Anderson film *The Life Aquatic with Steve Zissou*, the brewers developed a new pale ale recipe, calling it **Jaguar Shark**. For its first canned release, they hosted a free screening of the film and handed out red beanie hats to anyone who purchased a six-pack.

Burlington is known for being a progressive city, and its politicians are no exception. Senator Bernie Sanders, who made a presidential bid in 2016 (and is running again in 2020), was once the Queen City's mayor and still has many fans in Burlington. In a homage to him, Zero Gravity has released multiple batches of their **Bernie Weisse Beer**, "a slightly sour and forward-thinking Berliner Weisse" style of beer.

To flex their skills at brewing sour beers, Zero Gravity developed **Strawberry Moon**, an American sour ale brewed with lactobacillus, 2,400 pounds of Vermont-grown strawberries, and pilsner malt from Peterson Quality Malt Pilsner Malt in Monkton, Vermont.

THE BREWERS

With Brewmaster Paul Sayler at the helm, he is assisted by an incredibly talented team. Justin McCarthy, formerly of Magic Hat Brewing Company, is now the Head of Brewing Operations at the Pine Street location. McCarthy is joined by Steve Theoharides, an award-winning[164] former brewer at Harpoon Brewing Company.

THE BRAND

A discussion about Zero Gravity would be incomplete without mentioning their creative and ever-shifting branding and design. Sayler gives all credit in this realm to Matt Wilson, a co-founder of the brewery and Director of Sales & Marketing. "I'm not a 'brand' guy," Sayler admitted. "I resist 'brand.' But fortunately we have some wonderful people who get it and will remind me whenever I have an allergic reaction to 'brand.' Because…in craft brewing, what 'brand' really is is the essence of the brewer. What is their psychic nature or their creative bent or their persona, their bizarre persona, is the 'brand' in craft beer." Wilson guides the brand's style following that psychic thread laid by Sayler, Saxon and the rest of the brewing team, bringing images to life to tell the story of the beer.

QUEEN CITY BREWERY

703-B PINE STREET, BURLINGTON

"World-class beer without the jet lag."

Named for the city of Burlington, Queen City Brewery opened its doors on June 6, 2014, in a renovated warehouse space on Pine Street.

Before becoming a professional brewer, founder Paul Hale homebrewed for over thirty years, logging over four hundred batches. "I always said it was a hobby, but my wife said, 'No, it's an obsession.'"[165] Armed with a PhD in chemistry from Northwestern University, Hale cites his dedication to the scientific method as the foundation for his success as a homebrewer.

In the early 1990s, Hale and some friends visited England for a "research trip." In two weeks, they tasted over sixty different beers, keeping detailed notes. Hale cites a visit to Larkins Brewery in Kent, England, as the inspiration for his plan to someday start his own brewery.

PARTNERS IN BREWING

Hale decided to marry his love of brewing with his desire to preserve European brewing traditions. In 2012, he made a blog post to see if he could find partners for "brewing bigger batches." Within ten minutes, he was contacted by a homebrewing mate, Paul Held, who expressed interest.

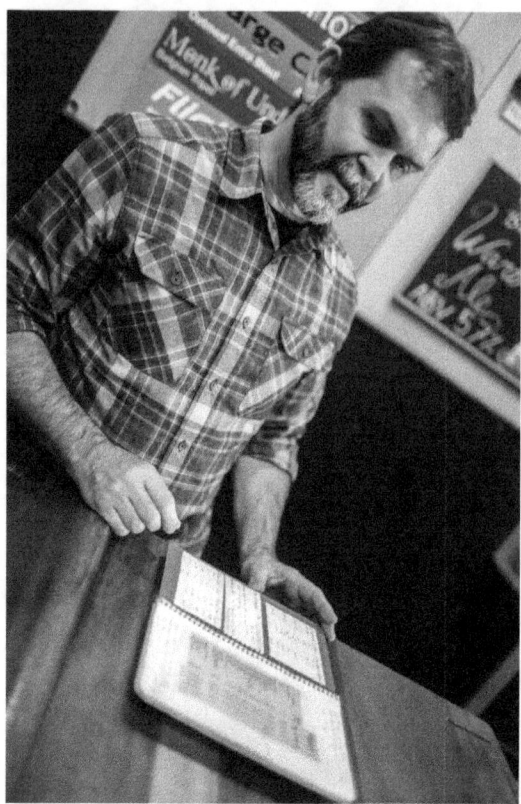

Left: Queen City brewmaster Paul Hale presenting some of his early homebrewing logs in his taproom. *Courtesy Paul Sarne.*

Below: In the Queen City Brewery taproom, an entire wall is occupied by vintage beer cans collected from all over the world. *Courtesy Paul Sarne.*

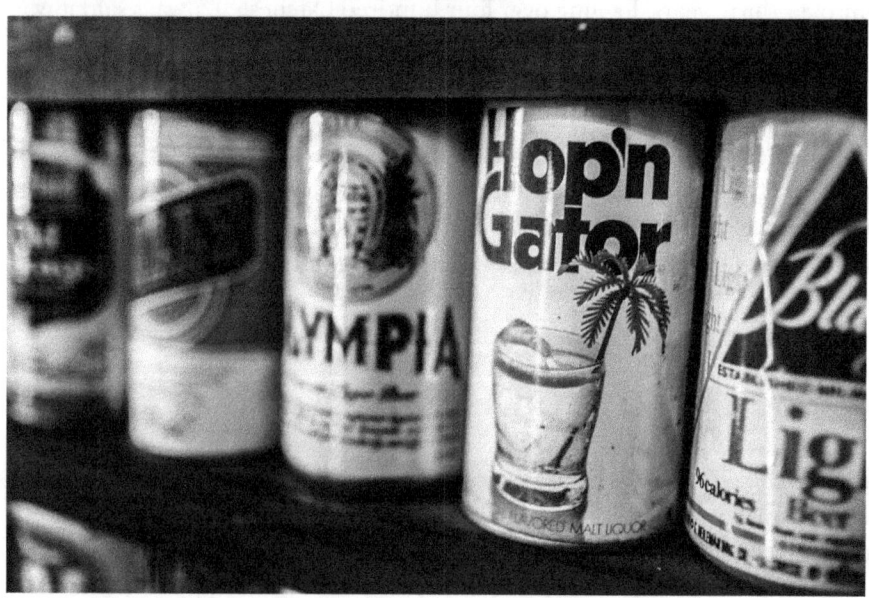

Another partner in the brewery project was Phil Kaszuba, who had been Hale's homebrewing partner for twenty years and had expertise in German styles. Having focused on perfecting processes like decoction mashing (no easy task for a homebrewer), Kaszuba's experience with these methods expanded the potential repertoire for Queen City Brewery.

Maarten and Sarah van Ryckevorsel designed the original brewery logo and all the marketing materials, including the tap handles, which are topped with a bright orange "Q."

Additionally, over twenty family members and friends became investors in the project.

EARLY BEERS

The first two batches of beer produced at Queen City Brewery were inspired by Belgian brewing traditions. **Saint Amandus**, named for the patron saint of brewers, was a Belgian-style amber ale. It was followed by **Duchess of Antwerp**, a Belgian-style blonde ale.

The next two batches celebrated British brewing traditions. Hale's visits to England were formative in his understanding of beer, and he drew inspiration from many of Britain's beers. **Landlady**, an homage to Timothy Taylor's Landlord Ale, was an extra special bitter brewed with English pale and crystal malts and hopped with Fuggles and Kent Goldings.

Landlady was quickly followed by **Yorkshire Porter**, a nod to the well-balanced traditional English porters rarely seen stateside. Yorkshire Porter quickly became the brewery's most successful beer and earned a year-round spot in the brewing rotation.

Many German-style batches followed, including **Munich Dunkel** and **City Beer**, a Kölsch-style ale.

THROWING STONES

On October 19, 2014, the brewers realized their long-term goal of brewing a commercial batch of **Steinbier** (literally "stone beer"). For over twenty years, Hale, Held, Kaszuba and their friend Rich "Monk" Evans have kept this Alpine beer style alive in their homebrewing circle. The style uses heated rocks to boil the wort.[166] Queen City Brewery used Graywacke rocks, a type of sandstone that is indigenous to Vermont and can withstand high heat and rapid cooling without shattering. The rocks were heated over a beechwood fire before being carefully lowered into the waiting wort. Due to the extreme temperature, the wort boils instantly and caramelizes around the rocks, lending smoky and caramel notes to the beer. After the rocks were removed and allowed to cool, they were later added to the conditioning tank, which allowed the caramelized smoky malt stuck to the surface of the rocks to dissolve into the beer. Steinbier has become an annual tradition for Queen City every October.[167]

BARTENDER: I'LL HAVE AN ARGUMENT!

Queen City was entering a mature craft beer market. In 2014, Burlington was already home to four breweries, with forty-six breweries in the entire state.[168] Not only was the market crowded, but Vermont had made a name for itself brewing IPAs. When Queen City opened, something was noticeably missing: hoppy beers.

"I didn't really even consider it," Hale said. His dream was to accurately make the beers that he was comfortable with: European styles, including those from Belgium, Germany and Britain. He focused on honing his techniques, like adjusting water chemistry and sourcing raw materials to recreate historical beers. His goal was to master and emulate his favorite styles and bring the flavors of Europe to Vermont.

Paul Sayler recognized Hale's dedication to mastering historical beer styles: "[We] are on the same path to developing and understanding beer. We both have a large degree of focus on beer styles and the history of brewing technique." Hale views it "as preserving heritage"[169] in an era when traditions are easily lost.

As IPAs topped the charts, Hale said that he basically missed the whole trend. "I was in my basement making all these other styles. So…I never even thought about it, really."

In early 2015, Queen City did finally produce an IPA—an English-style IPA, dubbing it **Argument** based on the fact that the owners argued about whether or not they would offer an IPA at all. Hale said the decision to make it an English-style IPA was because of his comfort with the English brewing tradition. He credits other Vermont brewers who excel at brewing hazy IPAs and questioned why he would ever try to compete with them in that space, stylistically speaking.

PAYING TRIBUTE

In August 2015, Queen City released a Scotch ale in homage to the late Greg Noonan, founder of Vermont Pub & Brewery. Over the previous twenty years, Hale had asked Noonan for advice on brewing Scotch ales. Noonan, who was a big supporter of homebrewing, was widely accepted as an expert on the style, as he literally wrote the book on it.[170] Noonan shared his recipe for Vermont Pub & Brewery's **Wee Heavy** with Hale, who brewed it at home for years, making small changes to the recipe. Queen City's **Gregarious** is a fitting tribute to the late master of Scotch ale.

NEW BREWERS

While entering their second year in 2016, the founders of Queen City realized that they needed to hire a brewer with commercial brewing experience. Lillian MacNamara was brought on due to her knowledge of industrial brewing equipment and commercial brewing processes.

MacNamara has brewing credits on both the West Coast and the East Coast. She started homebrewing after receiving dual degrees in anthropology and geology from SUNY Plattsburgh in 2002. After a successful first batch using all grain, she felt confident that she would be good at brewing.

She broke into the commercial brewing world by landing a job at Oregon Trail Brewery in Corvallis, Oregon. Digging up hop rhizomes, washing kegs and brewing root beer put her in the right place at the right time when the brewer quit. In 2011, she received a call from Mike Gerhart, who offered her a job at Otter Creek Brewing in Middlebury, Vermont. She hopped in her car and drove across the country only to end up living in a campground while Hurricane Irene ravaged Vermont. She weathered the storm and brewed at Otter Creek for three years. Job opportunities kept coming for MacNamara: In March 2014, she took a brewing job at Magic Hat, and in February 2015, she became head brewer for Hop'n Moose Brewing Company in Rutland, Vermont.

In July 2016, MacNamara was named head brewer at Queen City Brewery and was charged with starting a barrel-aging program in addition to developing new recipes.[171] Her first barrel-aged beer was **Old Monty**, an English-style barleywine aged in Chardonnay barrels from Premium Wine Group on Long Island. Next came a saison aged in a Cabernet Franc wine barrel, which was infected with an indigenous strain of Brettanomyces. The first release in the "Pint Street Barrel Project" was **Saison Aged in a Red Wine Puncheon with Indigenous Brett**.

In October 2016, Ben Gostanian, previously an assistant brewer at the South Burlington location of Farnham Ales & Lagers, joined the Queen City brewing team.

COOKING WITH BEER

Chef and cookbook author Sandi Earle released a book of recipes in late 2017 with each recipe featuring a Queen City beer. Here is a new recipe created by Earle that uses Queen City **Rauchbier**:

RAUCHBIER MAPLE BBQ MEATBALLS
Created by Sandi Earle

Rauchbier Maple BBQ Sauce
2 tablespoons canola oil
1 medium white onion, diced small
4 cloves garlic, peeled and chopped
6 sprigs oregano, washed, picked and chopped
2 cups Queen City Brewery Rauchbier
¼ cup Vermont Maple Syrup
2 cups ketchup
1 tablespoon chili powder
1 teaspoon onion powder
1 teaspoon garlic powder
1 teaspoon ground mustard
1 teaspoon cayenne
2 teaspoons kosher salt
1 tablespoon black pepper
2 ounces apple cider vinegar
2 teaspoons Worcestershire sauce

For sauce: Sauté onions in the oil for two minutes until the onions are translucent. Add garlic, oregano, Queen City Brewery Rauchbier, maple syrup and ketchup. Cook for five minutes. Add the remaining ingredients and simmer in a heavy-bottom stock pot on low heat for about two hours, stirring regularly.

Meatballs

1.5 pounds Vermont-raised ground beef
1 pound Vermont-raised ground pork
1 pound Vermont-raised ground rabbit
2 white onions, peeled, finely diced
6 garlic cloves, peeled, chopped fine
1 large shallot, peeled, finely diced
1 cup Vermont Bread Company 10 Grain breadcrumbs
3 eggs
2 teaspoons salt
1 teaspoon black pepper
1 teaspoon cayenne

For meatballs: Mix all ingredients together until well incorporated. Form into one-ounce balls, placing them on a sprayed sheet pan. Bake at 350 degrees Fahrenheit for 8 to 10 minutes. Then, in a four-quart saucepot, combine meatballs and Rauchbier Maple BBQ Sauce and simmer on low heat for about 40 minutes, stirring occasionally.

DOWNSIZING TO SCALE UP

As of 2017, Queen City Brewery planned to produce 1,200 barrels of beer per year. Initially, the beers were offered on draft and in twenty-two-ounce bomber bottles. Then, in early 2017, Queen City released downsized bottles of **Yorkshire Porter**, **Hefeweizen** and **South End Lager** in twelve-ounce six-packs of bottles, allowing them greater access to the retail market. At this time, breweries were experiencing a dip in bomber sales as drinkers shifted away from large-format single bottles toward smaller configurations. In 2019, Yorkshire Porter was released in twelve-ounce cans, with other canned offerings expected to follow later in the year.

As the bottle size got smaller, the footprint of the brewery grew. In 2018, Queen City expanded into the part of their building that abuts the street. The new space will be used for special events and to host live music.

SIDEBAR

The tasting room bar itself was purchased from the now-defunct Ethan Allen Club. The members-only social club operated for 114 years on College Street before closing in 2010.[172] Queen City bought the Ethan Allen Club mahogany bar, which dates to 1972. The bar was refurbished by Held, a skilled woodworker, before being installed at Queen City Brewery. Held's son, Jason "Metal Jay" Held, eventually became the tasting room manager for the brewery, serving beers over this reclaimed bar.

SIMPLE ROOTS BREWING

1127 NORTH AVENUE NO. 8, BURLINGTON

"Making memories one beer at a time."

Tucked into Burlington's New North End district is a brewery in an unlikely spot: a shopping plaza. On paper, the husband and wife behind it, Dan Ukolowicz and Kara Pawlusiak, are an unlikely duo to open a brewery in said unlikely spot: Pawlusiak is a middle school counselor, and Ukolowicz was a high school biology teacher.[173]

The two had a simple dream—to create a gathering spot for locals in a part of Burlington that is off the beaten path. The pair founded Simple Roots Brewing in September 2013. Pawlusiak had grown up in this neighborhood and wanted to put down roots in her hometown. After the two married, they started working on Ukolowicz's dream to be a professional brewer.

He had been homebrewing since 1997 and developed a love for it. As an academic scientist, Ukolowicz was methodical in his research. Before brewing a new style of beer, he would read all the relevant literature he could get his hands on. So, when he decided to go pro, he felt like he needed to study hard. "I really felt like I need a degree of some sort," he said.

As luck would have it, there was a world-renowned brewing school in Vermont: American Brewers Guild (ABG). ABG was founded by Steve Parkes and Christine McKeever and operates out of their brewery, Drop-In Brewing, in Middlebury, Vermont. Ukolowicz started on his new degree in early 2013.

After twenty-two weeks of online courses and one week in residence at Drop-In Brewing, the time came for Ukolowicz to secure an internship at a brewery. For him, the choice was crystal clear: it would be Zero Gravity Craft Brewery at American Flatbread Burlington Hearth.

Determined to work with Zero Gravity, and after seven unreturned phone calls, he managed to get a call back from brewmaster Paul Sayler, who agreed to an internship. Ukolowicz said he was drawn to Zero Gravity because of his love of beers that are brewed "true to style."

Being a schoolteacher has its advantages when trying to learn a new profession. As the two-month internship commenced during his 2013 summer vacation, Ukolowicz got to work with Sayler, then with assistant brewers Destiny Saxon and Justin McCarthy in the basement brewery at Burlington Hearth.

In the fall of 2013, with his Craftbrewers' Apprenticeship program completed, Ukolowicz started filing brewery permits. He and Pawlusiak started to renovate their garage into a professional brewery by early 2014.

The first professional brew was finished in their garage on May 2, 2014, and was appropriately named **American Dream**. The pre-Prohibition–style cream ale "tastes like beer," Ukolowicz joked during an interview, and is still one of the breweries' top five sellers. They attribute its success to the style's appeal to drinkers looking for light beers as well as craft beer aficionados in search of something unique. By brewing it with Chico yeast, pilsner malt and corn, Simple Roots aimed to make American Dream approachable and authentic to the historical style.

The next professional brew was **Gose The Destructor**, a slightly tart, German-style gose brewed with coriander and flaked sea salt. The name, an homage to the movie *Ghostbusters*, almost started a faux feud between Simple Roots and Zero Gravity. During his internship, Ukolowicz worked on a gose recipe with them and encouraged them to name it Gose The Destroyer, a riff on "Gozer the Gozerian." They did make one batch, but later dropped the name. When Simple Roots released its own gose, Ukolowicz urged them to bring back the name. "It'd be fun! We could talk shit about each other on Twitter!" But they weren't having it.

These first releases were only available in twenty-two-ounce bombers at the Burlington Farmers' Market on Saturdays during the summer. As business grew, they started to offer kegs to a select few local bars and restaurants, like La Boca Wood Fired Pizzeria and the Farmhouse Tap & Grill.

On Friday, October 10, 2014, the duo achieved another of their dreams: doing an event at the Farmhouse. "Having…kegs at Farmhouse was a dream come true," noted Ukolowicz.[174] The downstairs Parlor Bar was packed with friends and family of the brewers, and the event was heralded as a success.

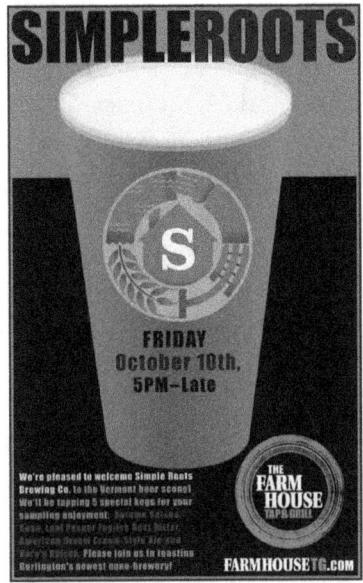

SIMPLEROOTS

FRIDAY
October 10th,
5PM-Late

We're pleased to welcome Simple Roots
Brewing Co. to the Vermont beer scene!
We'll be tapping 5 special kegs for your
sampling enjoyment: Maroon Saison,
Cana Leaf Pepper Jupiter Next Mister,
American Brown Cream-Style Ale and
Barry's Eclipse. Please join us in toasting
Burlington's newest nano-brewery!

THE FARM HOUSE TAP & GRILL

FARMHOUSETG.com

Poster advertising a Simple Roots Brewery tap takeover at the Farmhouse Tap & Grill. This was a "dream come true" for Simple Roots cofounder Dan Ukolowicz. *Courtesy Farmhouse Tap & Grill.*

As Simple Roots picked up steam, it was time to work on another dream—expanding the brewery into a permanent location that could welcome visitors and be a meeting place for the local community.

They selected an empty storefront in the Ethan Allen Shopping Center on North Avenue. There, the brewery would augment the current offerings, which included locally owned restaurants, a grocery store and 802 BWS (a large beer, wine and spirits shop). There was also a laundromat; folks could come to the brewery to grab a pint in between wash cycles: "Keep my tab open—I have to switch this load to the dryer!"

A week after the 2016 Vermont Brewers Festival, Simple Roots quietly opened their doors in the city's New North End. Pawlusiak is proud to serve beer to their neighbors. "The concept of having a neighborhood brewery is not just like, 'Oh, you know, that's just a cool thing.' Most of our customers are people in this neighborhood. We wanted to bring something else to this area for people to enjoy."

The flip side of choosing the New North End location for their brewery means that they're off the map for a lot of people. "While we were very intentional about picking this spot, we also knew we were taking a pretty serious risk," Pawlusiak said. "If you don't live out in the New North End, or you don't work here, you don't really come here. It is a destination." But they were rewarded for their bold choice and gained some national attention when *Imbibe Magazine* highlighted Simple Roots in a 2017 article about "under-the-radar beer destinations."[175]

With the new three-and-a-half-barrel brewery up and running, Simple Roots moved away from bottling in bombers, a package size that had rapidly declined in popularity with consumers, and started to can beers with a Twin Monkeys Yampa three-head canning unit, filling about twelve cans per minute. The first beers to be canned were **Citra & Amarillo Pale Ale, Good Neighbor IPA** and **Elderflower Saison**.

Sixteen-ounce cans allowed Simple Roots to tell a more robust story via beer labels. The duo worked with local graphic artist Jeff Holmes, who designed their logo and makes woodblock prints for each label. Thematically, Simple Roots achieved a unique Vermont identity by relying on the colors of Vermont's famous foliage season: oranges, reds and browns. The founders wanted to stay away from using green to represent Vermont, feeling as though too many other Vermont brewers were already utilizing *Green* Mountain State branding.

Like any Vermont brewery worth its salt, Simple Roots promotes the craft of beer and food pairings. Working with Vermont Creamery of Websterville, Vermont,[176] Ukolowicz and Pawlusiak host regular beer-and-cheese pairing events in their taproom. Pawlusiak recounted that one of her favorite pairings was the Vermont Creamery St. Albans, a gooey, soft-ripened, cow's-milk cheese made in the style of French St. Marcellin cheese, paired with **Citra & Amarillo Pale Ale**.

Even with so many goals accomplished, Ukolowicz and Pawlusiak aren't done dreaming yet. On deck is a plan to put their Polish-style pilsner, **Na Zdrowie**, into cans. "Man, that would be fricken awesome," Ukolowicz exclaimed in an interview.

FOAM BREWERS

112 LAKE STREET, BURLINGTON

"Foam. It's always been an unconscious calling, like drifting into a cumulus cloud bank on a sun-drenched day or collapsing in a frothy curl of a breaking wave. Foam—its natural beauty that rises up and is here and gone in a moment in time." [177]

On the shores of Lake Champlain, there is a sturdy brick building that has survived for over a century and a half. If the bricks could talk, they'd tell a rich story about the building's use, starting in the 1850s, when it was as a lumber processing and drying facility, then a door manufacturing plant in the 1870s, then an all-purpose specialty shop that made everything from leaded glass to blinds. [178] They would tell of fires that ravaged the waterfront district but left this building unharmed. Then came the offices and housing units, a salon and a restaurant space in the mid-1980s. [179]

When the last tenant of that restaurant space left in 2015, the space was on the market but seemed to be cursed. It was viewed as a "dead zone," according to Jon Farmer. [180]

A band of local brewers, however, saw a golden opportunity.

"There was a lot of naysayers about us coming down here," admitted brewmaster Todd Haire. If any business was going to survive in a "dead zone," a brewery with ample parking and a beautiful view of the lake would certainly be a winner.

Foam Brewers opened its doors on April 28, 2016, but it had taken this experienced band of brewers two years to bring the Foam to life.

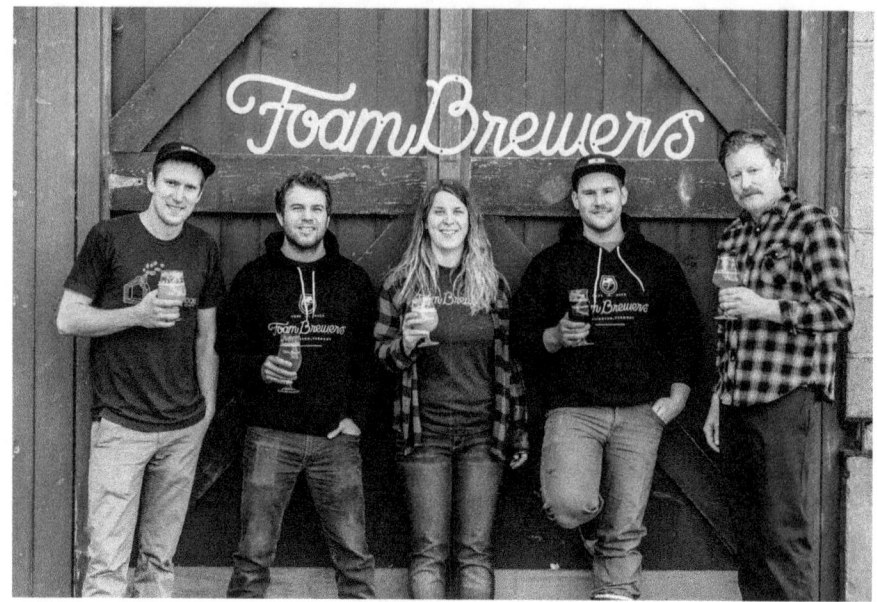

The founders of Foam Brewers (*left to right*): Jon Farmer, Sam Keane, Dani Casey, Robert Grim and Todd Haire. *Courtesy Jack Whitney.*

BUBBLING UP

In the mid-1990s, a homebrewer named Todd Haire was awarded the Michael Jackson Beer Education Scholarship and used it to attend the Siebel Institute. He earned his diploma in brewing alongside industry giants like Bradley Coors and Jennifer Yuengling.[181]

While on a short brewing stint in New York City, Haire met another brewer named Paul Sayler, and the two became good friends.[182] When Haire and his wife moved back to Vermont, he interviewed for a position at Magic Hat. Magic Hat cofounders Bob Johnson and Alan Newman held that interview in Burlington at the locally notorious (and now defunct) Chickenbone Café.

Haire was hired as a brewer, and during his thirteen-year tenure,[183] he solidified his place in the Burlington beer scene and befriended coworkers like Matt Cohen and Dani Casey. When Magic Hat was sold to NAB, Haire and Cohen decided it was time to move on. Cohen established his own brewery, Fiddlehead Brewing Company, and Haire joined Switchback.

During his four years at Switchback, Haire mentored up-and-coming brewers like Sam Keane and Robert Grim. Both Keane and Grim came with nutrition and food science degrees. Grim had brewed at Plattsburgh Brewing, and at Switchback, he was tasked with testing new recipes on the pilot system. Keane put his skills to work establishing a sensory training panel and brewing on the sixty-six-barrel brewhouse.

Foam Brewers started to form when Casey left Magic Hat to join Switchback as tasting room manager and when Jon Farmer was hired to work in the Switchback tasting room.[184]

QUIETLY FERMENTING

Grim and Keane approached Haire in 2015 to team up on their own brewery. However, they learned that Haire had simultaneously been working on a brewery business plan with Bill Mares.

Haire and Mares had known each other for some time; both were brewers, with Mares a homebrewer, and both were avid beekeepers. They had been quietly working on an update to Mare's book, *Making Beer*, and also on a business plan to open a blendery called House of Fermentology.

Haire decided the timing was right, and he agreed to join Mares on the blendery project as well as Grim and Keane on their brewery plan. Haire, Grim and Keane brought in Casey as general manager and Farmer as creative director, and the quintet dove into the foam.

FOAM IS "ARTISTIC FREEDOM"[185]

Designing a space with a "good feel" was central to their business plan.[186] Foam Brewers was designed to include local artists and artisans in the creation of a space for communal experiences.

To achieve this goal, all facets of the brewery were designed to engage patrons. "Everything is an interaction here," Grim explained. "One of the reasons the bar is laid out the way it is—it's not a straight line, you don't have to look down the table. You're always facing somebody."

"All the seating here encourages conversation and interactions," added Casey.

Haire sprang forth with an apropos beer pun, saying the features are "all nucleation sites!"

This curving concrete bar was hand-poured by Vermont Eco-Floors (Charlotte, Vermont) and is studded with inlays made by Mad River Glass Gallery (Waitsfield, Vermont) and stones straight from Lake Champlain.

The seven-barrel brewhouse sits in one corner of the space and is surrounded by custom fencing by local metal sculptor Kat Clear. Other decorative metal art installations, including the tap handles, were made by Christopher Solbert, formerly of Charlotte, Vermont. Outside the brewery is a large mural painted by Charlie Hudson,[187] who also created artwork for Foam's labels. And the creative force behind the entire buildout was the famous contractor Russ Bennett, whose other projects include the Magic Hat Artifactory and music festivals such as Bonnaroo.[188]

"It's all the little details that really matter to us," Grim said, while the other members of Foam nodded their heads. "And to have people experience a beer in the perfect environment that we see fit. All the way from the music that's on to walking through the archway with the lighting."[189]

Haire elaborated on how the name Foam fit the whole experience: "It's everything that we all believe that brewers strive for when they make great beer. It's the visual aspect of the crowning moment of beer. The first thing is your visual, and your second thing is your aromatics, then the bitterness of the foam and then the taste. It just transcends everything from the glass to your mind. We go to lengths to pour the beer the right way so visually it's what you want to see. No 'Iceman' pours! [Haire wagged his finger while laughing at this final line.]"[190]

LOCAL LIBATIONS

Vermont has a long history of supporting Vermont businesses first and warmly embracing local brewers the same way they do local farmers.[191] The brewers at Foam have made it their mission to embrace Vermont farmers.

Haire believes that the only way to improve the quality of local ingredients is for brewers to support these suppliers when they're just getting started. "These guys need an opportunity. Brewers should be open-minded and give it a try."

Vermont-grown hops are acquired whenever possible from local hop farmers and the University of Vermont Northwest Crops and Soil Program, which runs an experimental hop yard in Alburgh. A significant portion of Foam's malt comes from Peterson Quality Malt in Monkton,[192] which purchases its grain from Thornhill Farm in Greensboro. Coincidently (or perhaps not), Todd Hardie, who owns Thornhill, is also a beekeeper.

But they don't stop with traditional brewing ingredients. Some dark beers were brewed with coffee from Brio Coffeeworks, and a batch of kvass was brewed with sourdough bread from August First Bakery. Foam has even collaborated with Miss Weinerz, a Burlington-area bakery. Ren Weiner used spent grain from Foam to make doughnuts and glazed them with icing infused with Foam beers. Then, the doughnuts went back to Foam, where the brewers offered them paired with beers.[193]

And then, of course, there's Vermont's other famous product: cheese. Casey said, laughing, "We go through a lot of cheese. I mean, like, a lot!"

THE FROTHY SUDS

To get the beer to the tap was an all-hands-on-the-brew-deck effort. The quintet worked on the buildout during the day and brewed at night.

With seven beers on tap, they opened to the public on April 28, 2016. Two saisons were offered: **Saison de Foam** and **The Froth Saison**. **Built To Spill**, Foam's first double IPA, is still in rotation in 2018. **Easy Keasey** (American wheat ale), **Deep City** (pale ale), **Lupi Fresh** (IPA) and **Tranquil Pils** (pilsner) rounded out the opening list.

Developing new beers is a collective process at Foam, and they try to brew with seasonally appropriate ingredients—something most modern breweries forego.

A sour beer program was always part of the plan, and in December 2016, they bought a ten-barrel foeder and utilize it in a Solera style. When the beer is ready, they rack seven barrels off for packaging and top off the foeder with the next batch.

In 2017, Foam Brewers released its first cans of **Built to Spill**. Subsequent releases have included **Pavement** (double IPA), **The Nameless** (IPA), **Experimental Jet Set** (double IPA) and **Compact Disc** (pilsner).

Foam Brewers produced about one thousand barrels of beer in 2017, most of which was sold at the brewery with only a few kegs left for special events.

EARLY ACCOLADES

In a very short time, Foam Brewers has garnered a lot of attention both locally and internationally. Less than one year after opening, Foam received praise from both *BeerAdvocate* magazine and RateBeer.com, both of which named Foam Brewers one of the best new breweries in the world.[194]

On the local level, the Lake Champlain Chamber of Commerce awarded Foam Brewers the A. Wayne Roberts Entrepreneurial Spirit Award in 2017 for exhibiting "a promising future and strong commitment to our community."[195] In 2018, Foam Brewers was a finalist for "Best Craft Brewery" in the *Seven Days* Daysies Awards.[196]

VERMONT: LAND OF IPAs

"It's not just Vermont that loves Vermont IPAs." [197]

The brewing scene in Vermont is known for innovation and creativity, which gave rise to the black IPA in 1990. [198] But that's not the only style of beer created in Vermont that has garnered international attention. Between 2003 and 2010, some Vermont brewers started to formulate beers that would eventually define a new style of IPA: the Vermont IPA. [199]

It started with the Alchemist Pub & Brewery. The Alchemist became locally famous for their double IPA, **Heady Topper**. Now one of the highest rated double IPAs in the world, Heady Topper has been an exercise in perfection. Co-owner and brewmaster John Kimmich has constantly tweaked the recipe over the years, [200] but the yeast has always been the same: VPB1188. Nicknamed "Conan," this yeast was given to Kimmich by Greg Noonan, of Vermont Pub & Brewery (VPB), who had originally acquired it in England. English malts play a role in the flavor profile, as does leaving the beer unfiltered. Packed with bright, floral, citrusy hops, the resulting beer is cloudy, soft in texture and supremely hoppy.

Conor Giard, an early advocate for delineating the Vermont IPA as its own style, recalled his first experience with Heady Topper: "I remember drinking Heady Topper in the [Alchemist] brewpub for the first time, and it was a totally different experience. It may have been the freshness or it was all the late addition hops, but the nose was so much more expressive. For me, the reason I thought Vermont IPA should be its own separate category of IPA was because it was such a different experience from the previous IPAs and DIPAs that I had come across."[201]

In 2010, after returning from a stint brewing in Europe, Shaun e. Hill opened a brewery on his family's land in Greensboro in Vermont's remote Northeast Kingdom. Hill Farmstead Brewery would go on to be named "Best Brewery in the World" an astounding five years in a row by RateBeer.com[202] for their juicy, hazy, soft pale ales and IPAs. **Edward Pale Ale**, named for Hill's grandfather, is the brewery's flagship and is brewed with pale and caramel malts and hops such as Centennial, Chinook, Columbus, Simcoe and Warrior hops, and is left unfiltered.

Another early adopter of the term Vermont IPA, Sam King, said that when he first tasted Edward, "the whole game changed....You didn't just let it slide down your throat. It exploded like a supernova in every conceivable nook and cranny it could find, and coated the palate with a sticky, juicy sorbet of citrus and pine."[203]

Heady Topper and Edward were the "archetypal bookends"[204] to a new style: Vermont IPA. Vermont IPAs—and Vermont-*style* IPAs brewed elsewhere—have a bright golden hue and range from hazy to cloudy thanks to being left unfiltered and additions of wheat or oats. Aromatically, Vermont IPAs exhibit notes of citrus and tropical fruits thanks to heavy dry-hopping regimens with hops like Citra, Galaxy and Mosaic. The bitterness leans significantly lower than American IPAs thanks to the shift toward late-addition hopping.

It's possible this style might have originated at the same location as the black IPA. According to beer writer Ken Weaver, VPB reportedly brewed hazy IPAs as far back as the mid-1990s; Kimmich corroborated this in an 2016 interview.[205] One of the earliest uses of the term "Vermont IPA" can be attributed to Hill in a blog entry dated November 23, 2009. He referred to a beer he was drinking at VPB as "the original Vermont India Pale Ale."[206] After interviewing Hill, it was clear that he was not using the term then the way it is used today; however, he did note that he drew inspiration from VPB's IPAs, which he described as: "Soft. Smooth. Little bitterness. And uniquely its own."[207]

The Vermont IPA style caught on like wildfire, and soon, other Burlington breweries were producing hazy, soft, juicy IPAs of their own. On the Ides of March in 2013, Zero Gravity Craft Brewery brewed a single-hop IPA using Citra hops and a portion of wheat malt, calling it **Conehead IPA**. By 2013, Conehead had become Zero Gravity's top-selling beer.

Whereas unfiltered beer was once considered inferior (of course, some brewers still think this), Switchback Brewery may have set the stage for the hazy Vermont IPA. Since 2002, **Switchback Ale** has been unfiltered and has ranged from hazy to cloudy. In April 2015, Switchback released **Connector IPA**, an unfiltered IPA with late additions of Citra hops and heavy dry-hopping with Mosaic and Centennial hops.[208]

In late 2015, Brad Japhe argued in an article for Eater.com that "there's a new beer style in town, and it's from Vermont."[209] Although Japhe adopted the style name (as did the editors of *Craft Beer & Brewing* magazine),[210] many of the Vermont brewers who were producing these juicy, hazy IPAs were reluctant to adopt the nomenclature. When Kimmich was asked what he thought of the term, he responded, "I think it's bullshit."[211] In the Japhe article, Kimmich added, "Personally, I find it a little arrogant to try and claim that we do something so different that it deserves its own category." Even Garrett Oliver, famed brewmaster of Brooklyn Brewing, was vocally opposed to granting Vermont IPA its own style category.[212]

But more Vermont brewers jumped on the style. Foam Brewers opened in Burlington at a time when the hazy Vermont IPA was already all the rage, and one can often find multiple hazy, juicy, not-very-bitter IPAs on their menu. And, perhaps poking a little fun at the mild naming controversy, Frost Beer Works in Hinesburg produced a beer called "**Another VT IPA**." Long Trail Brewing, of Bridgewater Corners, released a juicy and hazy IPA under the name "**VT IPA**" in early 2018.

As the popularity of Vermont beer soared, brewers from around the globe started to use the phrase "Vermont IPA" on their labels, hoping to bank on some of Vermont's cachet. Examples turned up in California, Maryland and Europe.[213] The only thing Vermont-y about some of them was they claimed to use a yeast strain originally from the Green Mountain State.[214] In April 2017, Vermont state representative Sam Young and state senator Tim Ashe contacted the Vermont attorney general's office, but at the time of print, no solution had been found to prevent the use of the "Vermont brand" on products that are not sold within Vermont's state borders.

En vogue brewers in Massachusetts and other New England states started to brew this style of IPA, and the nomenclature shifted to New England IPA[215] or Northeast IPA. By mid-2017, *BeerAdvocate* founders Todd and Jason Alström declared the New England IPA to be an "official" style.[216] In 2018, the Brewers Association also added a new set of style guidelines under the IPA umbrella, choosing the phrase "Juicy or Hazy" to avoid a territorial debate like the one that had raged around the black IPA. Chris Swersey, of the Brewers Association, said that Charlie Papazian "is really thoughtful about how we name [beer styles], and unless it's truly associated with one place over time, like Vienna lager or Irish stout, we try not to pin a place on it." But the Brewers Association said it had to recognize the reality of the style. "These beers are here to stay," said Swersey. "It would have been a disservice to ignore that."[217]

While Vermont didn't get to name the style, it's clear that this style that has gained so much international attention started in the Green Mountain State.

HOUSE OF FERMENTOLOGY

777 PINE STREET, BURLINGTON

"Simple names, complex beers." [218]

Orange Dot. Pink Dot. Yellow Dot. The House of Fermentology produces complex, barrel-aged, wild-fermented beers and gives them decidedly simple names. It's an homage to sour beer blenderies of Belgium that identify different bottlings with a blaze of color instead of a batch number or fanciful name. [219] The idea is to bring the focus away from an oddball name and get drinkers to think about what's inside the bottle. [220]

House of Fermentology was dreamt up by Todd Haire and Bill Mares while they were working together on an updated version of Mares' book, *Making Beer.* The two had become fast friends while Haire was brewing at Magic Hat and Mares was teaching high school.

"I'd go there to pick up a growler or two [of beer] after school," said the now-retired Mares. "One of my former students was working there, and he said, 'You know, we've got a brewer here who's also doing bees.'" [221]

Turns out that beekeeping wasn't the only thing the two had in common. They were both longtime brewers, with Mares being a homebrewer. Additionally, both men were raised in Texas and have a history of wearing mustaches. [222]

As their friendship blossomed, they kept beehives together and talked about beer. In their book, Mares wrote that it struck him how similar beekeeping was to brewing beers with wild yeasts. "We're working with live beings on the cusp between the domestic and the wild. With both, we

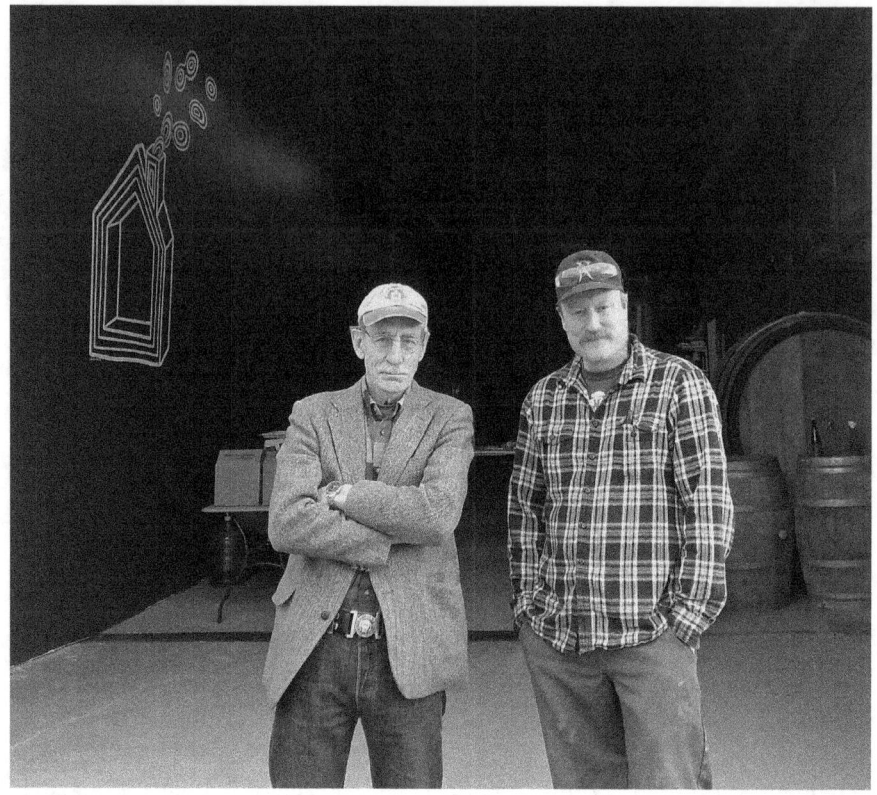

Bill Mares (*left*) and Todd Haire teamed up to open House of Fermentology, a wild ale blendery located on Pine Street. *Courtesy Bill Mares.*

are coaxing them to decide which course to take."[223] Haire once used their honey in a Braggot ale at Magic Hat.[224]

When Haire left Magic Hat to brew at Switchback, the two agreed to co-author the third edition of Mares' *Making Beer* book. They traveled around Vermont and even made a beer pilgrimage to Oregon together, researching the modern craft beer landscape for the new edition.

Inspired by the likes of de Garde Brewing, Cascade Brewing and The Ale Apothecary,[225] they solidified their decision to open a beer blendery instead of a full brewery. This meant less capital, less equipment and fewer requirements in terms of a physical space.[226]

But first, they needed a plan and a name. Haire had a penchant for fermenting foods like kimchi, pickles and kombucha in his house—so much so that he joked to Mares that "the name of his own home should become 'the House of Fermentology.'"[227] This turned out to be the perfect name for their project, which would be a study in wild fermentation in a simple, no-frills setting. After scouring the Burlington area, they found that simple setting: a bay in a tire shop.

During the planning stages for House of Fermentology, two other Switchback brewers approached Haire about teaming up with them on a new small brewery. Sam Keane and Robert Grim wanted to "size down" and create a brewpub.[228] Haire was already committed to the House of Fermentology project, but after reading their business plan and discussing it with his wife, Monica, he decided the two businesses could perfectly mesh together. What would eventually become Foam Brewers could supply the wort for House of Fermentology, and House of Fermentology could provide some unique beers to be sold at Foam Brewers.

The Vermont beer community is a tight-knit group, and brewers will often help new projects get off the ground. Shaun e. Hill (Hill Farmstead Brewery), Matt Cohen (Fiddlehead Brewing) and the team at Citizen Cider all helped Mares and Haire source used barrels, foeders, tanks and other equipment.[229] The first batch of wort for House of Fermentology was brewed at Fiddlehead, and then some batches were brewed at Zero Gravity while Foam Brewers was being built. All subsequent batches have been produced at Foam Brewers.

House of Fermentology is built on a house culture of wild yeasts and bacteria that behave less predictably than standard brewer's yeasts. In *Making Beer*, Haire made the analogy that this style of brewing is like jazz—there's a lot of improvisation, and every time they play a song or fill a barrel, it's going to be slightly different than any other version of itself.[230]

Used sixty-gallon wine barrels, former cognac puncheons and foeders of various sizes have been inoculated with this house culture and filled with beer at various stages of fermentation and maturation. Some barrels get fruits, herbs or—you guessed it!—honey added for secondary fermentation and extended aging.

Each release is blended by Haire and Mares, bottled in champagne-style half-bottles and designated with a colored dot. These bottles are sold at House of Fermentology and at Foam Brewers.

TAPROOMS

"A taproom is more than just a bar; more than just a place to grab a pint after work. A taproom is the cradle of beer culture, the place where craft beers from all over the globe first got attention....Anyone can build a great draught list, but the real taprooms build a culture."[231]

B
urlington's beer culture doesn't just revolve around breweries. For any vibrant beer city, the landscape must also include a robust collection of taprooms. While developing his business plan for American Flatbread Burlington Hearth, Paul Sayler noted that although the city had an established brewpub, Vermont Pub & Brewery, it lacked a bona fide taproom.

Sayler defined the taproom as providing a space for brewers and patrons to expand their knowledge of beer styles. "I view the role of the taproom as kind of that all essential dialectic that allows evolution."[232] To bring this culture to the Queen City, American Flatbread Burlington Hearth was founded as a "taproom with house beers." The draft list was split between house-brewed beers and the best beers that they could track down from around the globe via relationships with importers like the Shelton Brothers and Merchant du Vin.

For six years, Burlington Hearth operated as a gathering place for locals to experience exotic beers. But as the popularity of the house brand, Zero Gravity, increased, the owners decided it was time to open up another business to provide the next level of taproom experience.

Just a block north of St. Paul Street sat an empty building that once housed a McDonald's. The new taproom would become one of Burlington's most successful restaurants.

THE FARMHOUSE TAP & GRILL

160 Bank Street, Burlington

The owners of Burlington Hearth teamed up with Jed Davis to open a new concept in Burlington's downtown: a farm-to-table restaurant featuring Vermont's finest meats and produce with a world-class taproom. A seventh-generation Vermonter, Davis had spent his career honing his skills at some of the country's top restaurants, including Le Cirque, Daniel and Union Square Cafe.

The Farmhouse Tap & Grill opened in May 2010. Davis summed up the mission thusly: "The menu is a vehicle for showcasing Vermont-grown, -raised and -produced ingredients. We start with what is available locally, and we build our menu foundations from there."[233] Chef-partner Phillip Clayton designed

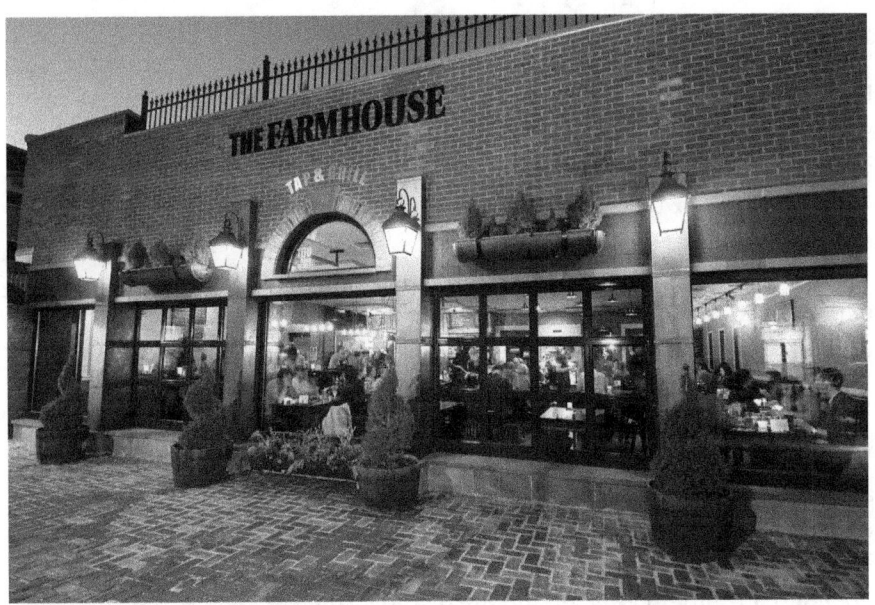

The Farmhouse Tap & Grill has become the most awarded craft beer taproom in Burlington in addition to garnering many accolades for its farm-to-table cuisine. *Courtesy Farmhouse Tap & Grill.*

the menu and built lasting relationships with Burlington-area farmers. Today, executive chef Kevin Sprouse continues this culinary mission.

The restaurant itself has deep connections to Vermont, and the owners wanted to utilize local products whenever possible. The bar in the dining room was crafted from Vermont Verde Antique serpentine stone from a quarry in Barre, Vermont. With a keen sense of humor, the owners decided to repair the existing faux brick floor from when the space was a McDonald's. They also placed Vermont slate chalkboards above the bar, ever so slightly slanted, to be reminiscent of a fast-food menu board. A "drive-through" window was installed at the end of the bar, although this particular McDonald's never offered drive-through service. Other Vermont flourishes included lighting and decorative fixtures fabricated by the highly acclaimed Conant Metal & Light shop on Pine Street, as well as tables made from Vermont-grown woods.

Sayler played a large part in setting up the foundation for the Farmhouse to take over the taproom responsibilities from Burlington Hearth. Chad Rich, a longtime staple behind the bar at Burlington Hearth, was named bar manager for the new project and built a robust beer program for the Farmhouse.

Rich dedicated one-third of the twenty-four draft lines to Vermont brews, while the rest were populated by the best beers from around the world. "Our taproom features Vermont's finest beers presented in conversation with the finest beers from around the world," was the mission statement of the beer program.

Food pairing was a focus for Rich as he worked to tightly tie the taproom with the culinary mission. During beer events, he would work closely with Chef Clayton to pair featured beers with Clayton's menu to provide an elevated guest experience.[234]

Six months in, the Farmhouse opened a downstairs speakeasy-style bar, dubbed The Parlor, complete with a reclaimed vintage bar, brick fireplace and decor from an old Vermont barn and a defunct Northeast Kingdom church.[235]

The old-school ornate bar itself came from a shuttered watering hole in Winooski, Vermont, called 38 Main Street Pub,[236] and was refurbished by artisan Michael Kuk. Then, in the summer of 2011, a seasonal outdoor beer garden was introduced.

When Rich left in November 2011 to open his own restaurant and brewpub, Prohibition Pig, on the site of the former Alchemist Brewpub, Jeff S. Baker II was tapped to take the helm of the beer program and guide it into its next phase.

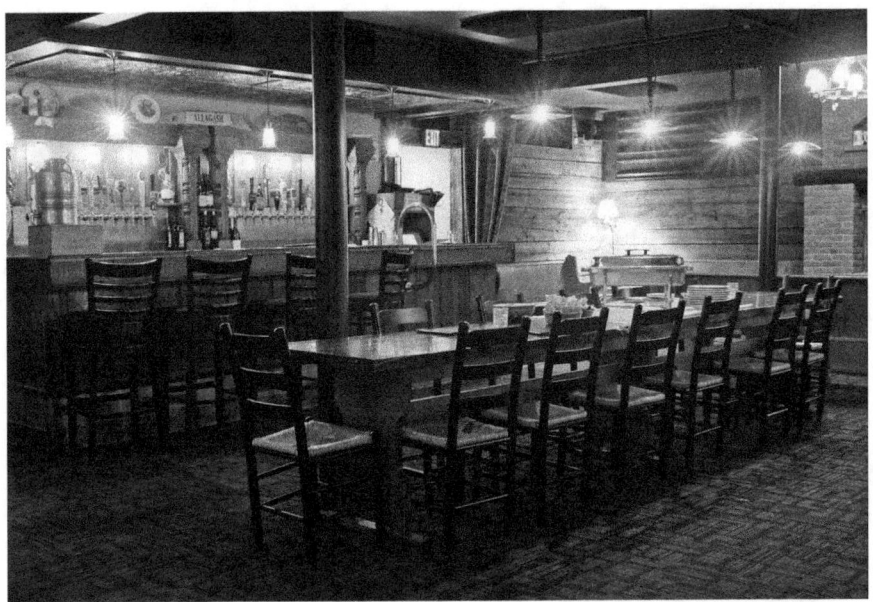

The downstairs Parlor Bar at the Farmhouse Tap & Grill has the feel of a friend's cozy basement, complete with a wood fireplace. *Courtesy Farmhouse Tap & Grill.*

"Great beer programs are really built on bottle lists," Baker said. "Anyone can have a great draft list; that's the easy part." Baker expanded the bottle list to two hundred offerings and ramped up a cellar-aging program for ageable beers like Trappist ales, barleywines and more in both bottles and kegs. The size of the beer list presented a challenge, however. "Most of what we did as a staff was to provide beer education to our guests and help to nurture their curiosity in a welcoming, judgment-free environment. Too many other taprooms take a snobby approach."

Before Baker left the Farmhouse in 2016 to lead the beer, cider and wine education program for a Vermont beverage distributor, he installed six additional draft lines. Justin Gould, who was previously general manager for the Farmhouse, was selected to take over managing the bar. Now, with thirty draft lines and a cellar full of aging beer, Gould pushed the beer program into its next chapter.

Gould believed in educating his staff and patrons through tasting and exploring different beers. "The more beer we can try, the more knowledgeable we are as a community."[237]

In addition to focusing on product knowledge, the Farmhouse is also a vocal proponent of proper beer service. "Draft line maintenance,"

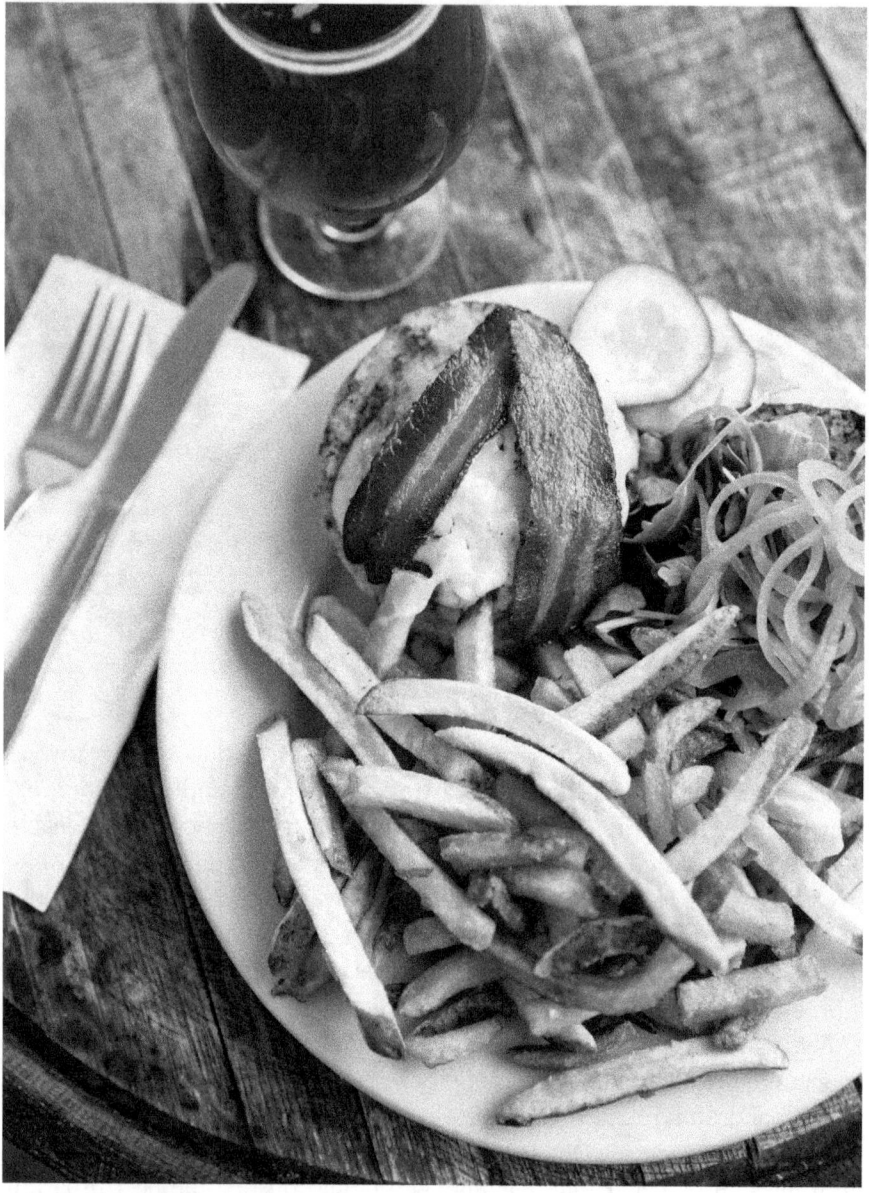

Voted "Best Burger in Vermont" eight years in a row in the Seven Daysies awards, the burger at Farmhouse Tap & Grill is made using local beef and fresh ingredients from Vermont farms. *Courtesy Farmhouse Tap & Grill.*

according to Gould, "is one of the most overlooked and important aspects of the bar side of the business." He believes bartenders are the stewards of the beer, and it's the bartender's "job to make sure that we are showcasing each beer as it was intended by the brewery....I am not going to serve you a beer through lines that are full of yeast and bacteria that have grown on them or faucets that have not been properly cleaned."[238]

In April 2018, Gould left for other opportunities in the Vermont beer industry, and Jim Hsieh, previously bar manager at Bluebird Barbecue in Burlington, took over the program.

Between 2010 and 2018, the Farmhouse earned many awards for both its beverage program and culinary prowess. Accolades include "Best Burger in Vermont" (for nine years in a row), "Best Bar" (in 2014) and "Best Draft Beer List" every year from 2014 through 2018 in the *Seven Days* Daysies Awards. CraftBeer.com acknowledged the Farmhouse on its list of Great American Beer Bars in 2013, 2014 and 2016. And *BeerAdvocate* featured the Farmhouse in issues No. 80 and No. 94, as did *Craft Beer & Brewing* (in 2014) and *Draft Magazine* (in 2015).[239]

A discussion about the Farmhouse would be incomplete without mentioning that many in Burlington consider the building to be haunted.[240] Before the McDonald's was built, there had been a lethal fire on the property in the Black Cat Café & Seafood early in the morning on May 1, 1977. An explosion resulted in the deaths of the owners' two children. Both owners were in the building at the time of the explosion. Mary Pappas survived, but John G. Pappas, her husband, was never located, and he was declared dead by state investigators. Former employees of both the McDonald's and the Farmhouse have reported seeing objects fly off shelves in the basement and hearing screams at early hours in the morning.

Burlington is also now home to other taprooms that offer a variety of unique experiences, as listed below.

THE ARCHIVES

191 College Street, Burlington

Founded by Matt Walters, Nate Beaman, Adam Lukens and Dan Chahine, the Archives is a unique concept for Burlington that combines classic arcade games with an upscale craft beer and cocktail program. The twenty-four draft lines, bottle list and cocktail program are managed by Sean McKenzie, an alumnus of the Farmhouse Group. McKenzie continues to push the envelope by innovating in the cocktail realm and curating a constantly rotating draft program that features both high-end and heritage beer brands.

FINNIGAN'S PUB

205 College Street, Burlington

Beer writer Allen McDuffee perfectly summed up the experience at Finnigan's Pub: "Craft beer taste, dive bar sensibility: Finnigan's is where a Lawson's Finest Liquids pint sits on the bar next to a Pabst, and nobody thinks twice about it."[241] Owner Terry Suskin has run this straightforward bar—which features flat-screen televisions, a pinball machine, a pool table and darts—for over ten years.

THE GARAGE

30 Main Street, Burlington

Offering twenty-two draft lines of Vermont beers and selections from abroad, plus a host of games such as shuffleboard and darts, the Garage has become a popular after-work hangout.

BEER LOVES CHEESE

In celebration of the fact that taproom culture combines craft beer and craft food, here is the recipe for the Farmhouse Tap & Grill's highly acclaimed Cheddar Ale Soup.

THE FARMHOUSE TAP & GRILL'S CHEDDAR ALE SOUP
by Chef Phillip Clayton

2 cups diced mirepoix (equal parts carrots, onions and celery)
2 cups peeled and diced potatoes
2 tablespoons Vermont butter
1 cup malty Vermont Ale (avoid hoppy beers)
1 cup stock (vegetable or chicken)
½ teaspoon cayenne
1 teaspoon sherry vinegar
2 cups milk
1 cup heavy cream
1 ¼ pounds Vermont cheddar, shredded
Croutons for garnish

Cook mirepoix in butter until tender. Add remaining ingredients except for cheddar, milk and cream. Simmer until potatoes are soft. Add milk and cream and heat to 190 degrees F. Pour into a blender, set to medium speed and slowly incorporate the cheddar. Add salt to taste. Add more sherry vinegar or cayenne as desired. Pass soup through a fine mesh strainer for even texture. Serve hot and garnish with croutons.

PART IV

BEER CULTURE IN THE QUEEN CITY

PINT STREET

Located two blocks west of Burlington's Church Street Marketplace, Pine Street runs north–south from the southern edge of the city's Old North End to the border with South Burlington. "Not too far back, Burlington's Pine Street was a desolate industrial corridor, with no good reason to visit the city's southern end," noted beer journalist Joshua M. Bernstein.[242]

Today, Pine Street has become home to a slew of beverage producers, artists and hip restaurateurs, leading to locals referring to it as "Pint" Street.

Johannah Leddy Donovan, who represents the Pine Street district in the state legislature, commented in a 2014 *Burlington Free Press* article that she was "excited" by all the changes to the area and that "there's a vibrancy I could never imagine." Donovan went on to remark about how Pine Street seemed to evolve so quickly. "It seems to me as though it just happened. Young people have made it happen....We've become a destination."[243]

The person who coined the term "Pint Street" was Paul Hale, founder of Queen City Brewery. When Queen City was awarded a license allowing for the service of full pints on-site, Hale declared that "Pine Street is now Pint Street" on his blog and the brewery's Twitter account on June 10, 2015.[244]

From north to south, the fermentories on Pint Street include Citizen Cider, Queen City Brewery, Zero Gravity Craft Brewery's production facility and House of Fermentology. At the north end, some consider Vermont Pub & Brewery and American Flatbread Burlington Hearth to be a part of Pint Street, despite both being one block over on St. Paul Street. Heading south on Pine Street, if you turn West on Flynn Avenue, you'll find Switchback Brewing, which is generally regarded as the southern end of Pint Street despite also not being on Pine Street proper.

In addition, Pint Street is also home to other libation purveyors, including a wine bar, coffee roasters, excellent restaurants and more.

CITIZEN CIDER

316 Pine Street, #114, Burlington

Before April 2014, before this road had become known as "Pint Street," if you stood in front of 316 Pine Street, you'd see an aging U-Haul rental location, a self-storage facility and a bustling post office. But that month, Citizen Cider moved its operation from the original location in Essex Junction to this U-Haul/storage facility space.

Cofounder Kris Nelson said that when they ran out of space in their original location, moving to Burlington's Pine Street just made sense. "We were looking for a place [where] we could build culture around cider, and Pine Street was ripe to be that spot. Burlington already had a thriving beer and food culture throughout the city and beyond," and by "plopping its manufacturing and world headquarters right at the headwaters of the South End Business and Arts District, [Citizen Cider would] certainly up the game of what the South End and Pine Street could offer the community."

The spacious cidery taproom offers tasting flights, full pours of both house-made ciders and Vermont-brewed beers and a robust food menu.

Justin Heilenbach, Bryan Holmes and Kris Nelson, the managing and founding owners, have since built a reputation on the East Coast for making authentic craft cider from Vermont and New York apples purchased directly from farmers and fermented without the use of added sugar or concentrates.

"Now, Pine Street has expanded its offerings and become a mecca of so much more," noted Nelson. "I am not saying we are responsible for making Pine Street cool, but we were a significant piece of the puzzle."

DEDALUS WINE SHOP, MARKET & WINE BAR

388 Pine Street, Burlington

For anyone cruising up and down Pine Street with a love of good food and drink, this is a must. One half of the space is a wine bar walled in with antique windows. The other half is a wine shop and market offering small-production wines and an assemblage of award-winning cheeses, charcuterie and dry goods from around the globe.

Dedalus has helped introduce Burlington to the world of artisanal wines with a focus on wines from uncommon areas, varietals and those created via unique production methods. With an extremely knowledgeable staff and bottles at all price ranges, there is no real need to explain why it has become so popular. It also helped that former in-house cheesemonger Rory Stamp was crowned the prestigious Cheesemonger of the Year at the 2018 Cheesemonger Invitational in San Francisco.

THE GREAT NORTHERN

716 Pine Street, Burlington

The group behind Zero Gravity teamed up with local culinary legend Frank Pace and his wife, Marnie Long, and took over the rest of the building at 716 Pine Street. Sharing a wall with the Zero Gravity production brewery and facing Queen City Brewery across the street, this is a prime location for a restaurant, and Pace expected it would be an immediate success.

The Great Northern opened its doors on May 14, 2017. Burlington-based food and beer writer Sally Pollak commented about the remodel: "The place is something of a man-made wonder: an industrial-size room softened by architectural features such as interior clapboard siding and a wooden arch flanking a trio of mirrors. A mega leather banquette stands before a stone fireplace, and a cast-iron lighting fixture hangs from patterned beams."[245]

The wooden arch over the trio of mirrors Pollak mentioned is a massive vintage backbar dating back to 1889.[246] The backbar was in service for 120 years in Seattle's J&M Café and Cardroom until bankruptcy led to it being put on the market. The owners of American Flatbread Burlington Hearth purchased the bar in 2009 and originally intended for it to be installed in a

Portland, Oregon, American Flatbread location, which never materialized. After sitting in storage for years, the glorious bar was refurbished and installed at the Great Northern by Vermont craft-builder Michael Kuk, whose other beer-related credits include design work for Burlington Hearth and The Farmhouse Group. According to Kuk, this bar had served many famous patrons over the years at J&M, including Wyatt Earp, John Wayne, Jimi Hendrix and Kurt Cobain.[247]

The cuisine at the Great Northern strikes a balance between local meats, seafood and vegetarian fare and combines aspects of Scandinavian and Asian traditions, utilizing pickling, fermenting and a raw bar. The beverage program, run by Jeff Baumann, is a highly cultivated endeavor that is sure to please both the casual drinker and the aficionado.

NOT JUST BEER

Craft beer tourists visiting Burlington will likely feel right at home at many of Pine Street's non–beer focused establishments. Here is a short list of other noteworthy attractions:

SOUTH END ART HOP: Held annually during early September and sponsored by South End Arts and Business Association (SEABA), this "art crawl" floods Pine Street with art exhibitions, concerts and performances. The year 2017 marked the twenty-fifth anniversary of the Art Hop. Magic Hat Brewery produces a beer in collaboration with SEABA called **Art Hop Ale**. seaba.com

ARTSRIOT (400 PINE STREET): A unique venue that offers live music, an art gallery and creative cuisines, cocktails and craft beer. artsriot.com

LAKE CHAMPLAIN CHOCOLATES FACTORY STORE (750 PINE STREET): Offering confections, bean-to-bar chocolates, ice cream, tours and, of course, free samples, LCC has produced collaboration chocolates with Citizen Cider and made hop-infused chocolates for the Vermont Brewers Festival. lakechamplainchocolates.com

FELDMANS' BAGELS (660 PINE STREET): Keep an eye on their marquee—they frequently bake fresh bagels using beer from Zero Gravity Craft Brewery, which is located two doors down. feldmansbagels.com

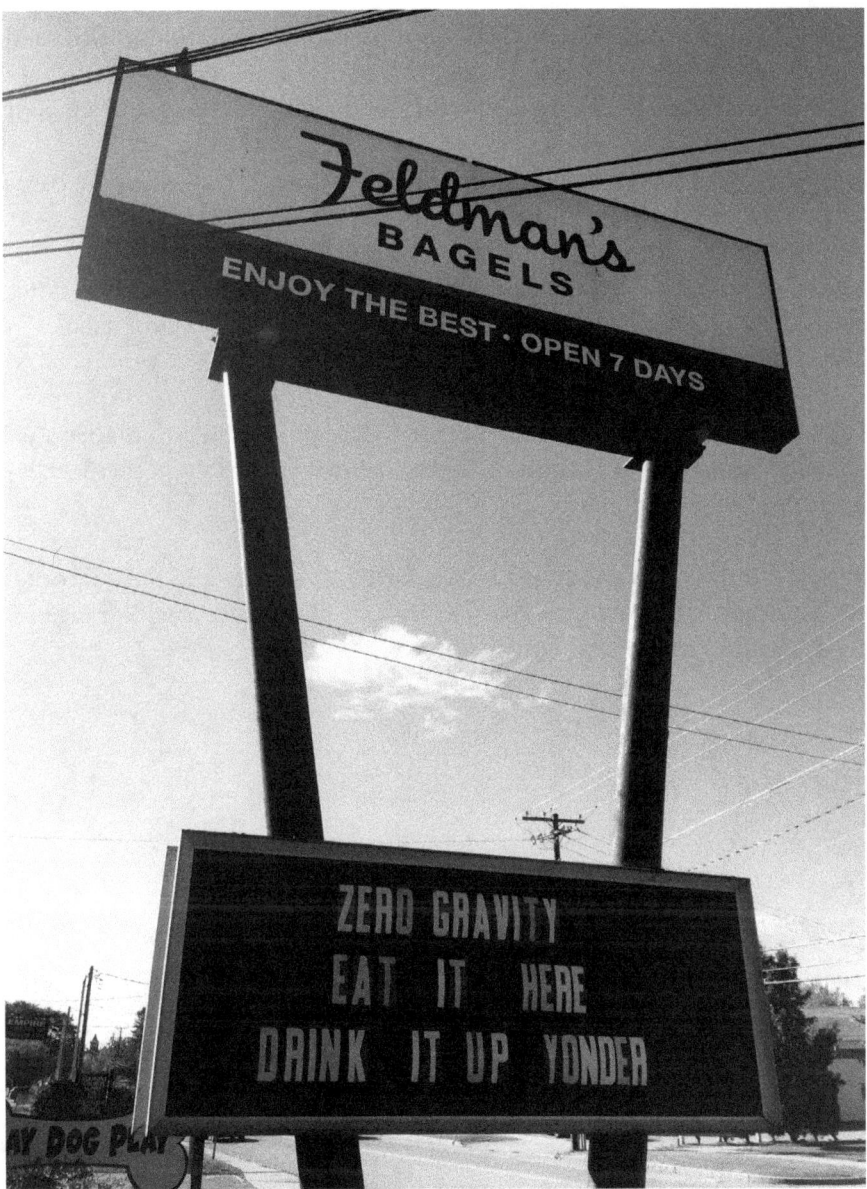

On "Pint Street," many businesses collaborate with each other. Feldman's Bagels regularly offers bagels made with Zero Gravity beer; the Zero Gravity brewery is only a few steps south on Pine Street. *Courtesy Jeff S. Baker II.*

BURLINGTON RECORD PLANT (660 PINE STREET NO. 5): A family-owned custom vinyl record–pressing plant. The music scene and the craft beer scene in Burlington are dedicated to each other. burlingtonrecordplant.com

SPEAKING VOLUMES (7 MARBLE AVENUE): A gallery of books, vinyl records and art. speakingvolumesvt.com

BRIO COFFEEWORKS (266 PINE STREET, SUITE 116): Specialty coffee roaster that works with local breweries on coffee beer collaborations and dabbles in barrel-aging coffee beans. briocoffeeworks.com

CITY MARKET SOUTH END (207 FLYNN AVENUE): Local co-op market that boasts an impressive collection of Vermont-brewed beers and other locally-made products. citymarket.coop

SPEEDER & EARL'S COFFEE (412 PINE STREET): A longtime local favorite, Speeder & Earl's is a coffee roaster and café that offers a wide range of coffees by the cup or the pound.

QUEEN (CITY) OF TOURISM

"Craft beer has become a way of life in the Queen City." [248]

B urlington's reputation as a craft beer and culinary mecca has spread far and wide, drawing in hundreds of thousands of visitors each year. "Vermont's tourism industry in particular has significantly benefited from the craft beer industry, as rural breweries become travel destinations for visitors from across the globe," according to Steven Cook, Deputy Commissioner of Vermont Department of Tourism & Marketing.[249]

Publications ranging from beverage magazines like *Craft Beer & Brewing*, *BeerAdvocate* and *Imbibe* to newspapers like the *New York Times* and culinary websites like Serious Eats have all published articles about Burlington.[250] This chapter is a celebration of some local voices that helped cultivate and promote Burlington's reputation.

BEER AND FOOD MECCA

Whether it is gracing the pages of magazines of *Food & Wine* or *Bon Appétit*, Burlington has emerged in recent years as a national tastemaker and trendsetter. Tommy Noonan, formerly of VPB, noted some years back that "many come here [to Burlington] for the beer, but no matter what they come for, they find other things, such as the food and dining experience."[251]

A standout example is the Burlington location of the Hen of the Wood Restaurant. Chef Jordan Ware has been nominated for the coveted James Beard Award for Best Chef Northeast, adding to the numerous nominations that chef/owner Eric Warnstedt has accumulated at the flagship location in Waterbury. The restaurant proudly supports and pours local Vermont craft beer with their award-winning cuisine alongside a stellar wine cellar and cocktail program.

Nearly every restaurant you walk into in Burlington has grown to take pride in their offerings on draft or by the can or bottle. No matter the cuisine or price point, craft beer has become entirely entwined with the local food scene.

BURLINGTON'S BEER CONCIERGE

Hotel Vermont opened its doors at 41 Cherry Street in May 2013. After a year of answering questions about Vermont beer, they decided to establish a "beer concierge" position. Matt Canning became the country's first concierge dedicated to beer tourism and has developed a rich catalog of options for hotel guests to explore the area's brewing scene.[252]

Canning tailors offerings to each guest's needs, "whether that's just having a beer at the bar and talking shop with someone or answering e-mails about where to find certain beers on certain days or actually doing [private] tastings. We recognize the fact that a huge amount of people walking through our doors are walking through our doors specifically with the intent of traveling here to drink beer and buy beer."[253]

Hotel Vermont offers two-day beer tour packages to guests throughout the summer and fall.[254] The tour starts with a beer dinner prepared by chef Doug Paine and paired with beers curated by Canning. The next day, guests travel with Canning throughout the Green Mountain State, grabbing lunch at a local farm, and the tour ends at a bottle shop so guests can stock up before leaving the state.

For Canning, it's a joy, not a job. "At the end of the day, we're talking about beer and Vermont, and that's awesome. You almost realize the beer is the by-product, and what's really important is showing people around our state…building relationships with people, talking about beer, explaining our culture here are the really important aspects of beer tourism."

BEER FESTIVALS

Burlington is home to a number of annual craft beer–focused events.

VERMONT BREWERS FESTIVAL BURLINGTON: This fest is held annually in the third week of July at Burlington's Waterfront Park and hosted by the Vermont Brewers Association, "a nonprofit membership organization that was founded in 1995 to promote and strengthen the culture of craft brewing in Vermont through marketing, education and advocacy for Vermont-made beer."[255] Thousands of attendees travel to Burlington for this festival, and in years past it has sold out in as few as eleven minutes.[256] In 2018, everyone celebrated the festival's twenty-sixth year.

OKTOBERFEST VERMONT: Created by Lou DiMasi, Oktoberfest Vermont provides a Bavarian-themed experience at Waterfront Park toward the end of September each year. The festival is expertly run by Red Clover Events and DiMasi's right-hand man, two-time "Best Bartender" Chris Maloney.[257] The festival's fourth annual event was held in 2018.

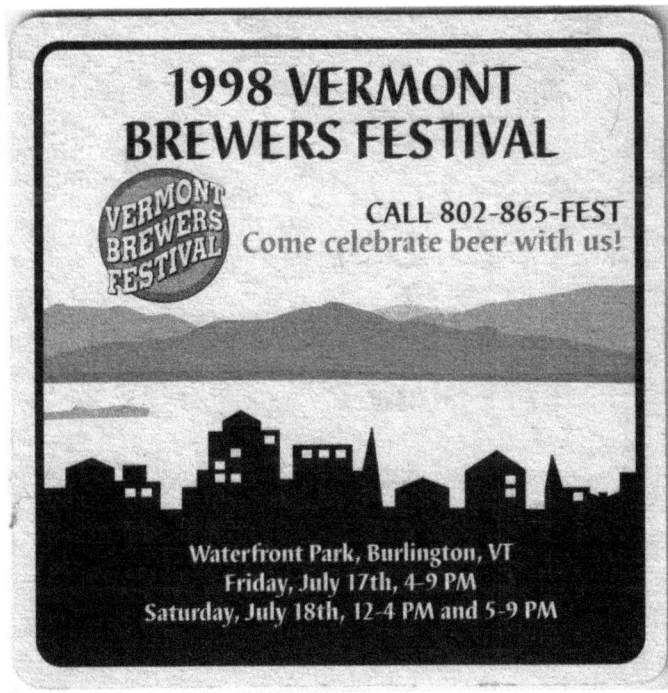

In 1998, the Vermont Brewers Association promoted their Vermont Brewers Festival via coasters. Today, the festival typically sells out well before the actual event. *Courtesy Kurt Staudter.*

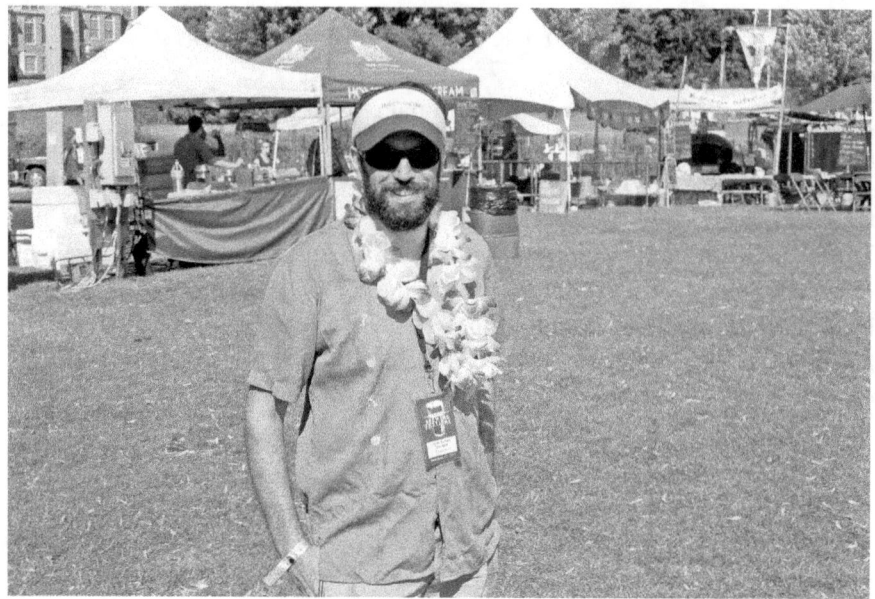

Matt "Matty O" Cohen, now owner and brewmaster of Fiddlehead Brewing Company in Shelburne, was previously a brewer at Magic Hat Brewing Company. *Courtesy Kurt Staudter.*

CITY BREW TOURS

For those with limited time in Burlington who want to visit three or four breweries in just four hours, City Brew Tours offers a handful of tour packages. Founded by Chad Browski in 2008, City Brew Tours has since expanded to other beer-soaked cities, like Boston and Philadelphia.[258]

BEER SCHOOL

The University of Vermont, which overlooks Burlington from its lofty hill, established the Business of Craft Beer program in February 2016.[259] Rather than being a class on *how* to brew beer, this online program is focused on the business side of the industry. The courses are taught by industry experts like brewery owners, marketing managers and distributing executives.

BEER JOURNALISM

As Burlington's reputation has grown, local journalists took note, starting projects centered on suds.

Brittney Hibbs-Kelson, producer of *What's on Tap?*, has been covering the "beer beat" for WVNY and WFFF since 2013. "People travel to Burlington from all over the world just to come and drink the beer we have here," Hibbs-Kelson said. "Craft beer has become a way of life in the Queen City, a way to connect with people you may have just met. Rather than, 'How's the weather out there?,' you are meeting people for the first time and saying 'Hey, have you checked out this IPA yet?'"[260] In November 2014, Vermont Pub & Brewery brewed a hibiscus wheat ale called **Hibbs Don't Lie** in honor of her contributions to Vermont beer culture.

Other media outlets also devote time and space to covering the Burlington beer scene. *Burlington Free Press* introduced a biweekly column called "Hops & Barley" in 2012 and the podcast *It's The Beer Talking* in July 2016; the podcast released sixty-seven episodes, garnered over twenty thousand downloads and had listeners in forty countries but went on hiatus in mid-2018.[261]

BURLINGTON'S EPONYMOUS BREWERIES

At this point, some readers of this book may be questioning the lack of inclusion of any modern breweries using the name "Burlington." The reason is simple—while there have been such breweries, none have existed within the Burlington city limits since Burlington Brewery closed around 1885.

Burlington Brewing Company, which opened in 1998, was located in the town of Shelburne, just south of Burlington, on Vermont Route 7. It wasn't a brewery or brewpub but rather a brew-your-own shop.[262] It closed in 2004.

In May 2014, Burlington Beer Company opened its doors in the town of Williston, located southeast of Burlington.

NOTES

The Early Days

1. Hamilton Child, *Gazetteer and Business Directory of Chittenden County, Vermont For 1882–83* (Syracuse: self-published, 1882), 113.
2. Ibid.
3. *Vermont Centinel*, April 5, 1805, 3.
4. *Burlington Free Press*, March 14, 1831, 2.
5. *Burlington Weekly Free Press*, April 4, 1828, 4.
6. *Burlington Weekly Free Press*, October 24, 1828, 2.
7. *Burlington Weekly Free Press*, January 6, 1832, 4.
8. *Burlington Weekly Free Press*, April 19, 1833, 4.
9. *Burlington Weekly Free Press*, May 23, 1834, 4.
10. *Burlington Weekly Free Press*, May 25, 1835, 3.
11. Abby Hemenway, the *Vermont Historical Gazetteer*, vol. I (Burlington, VT), 1867, 706.
12. *Burlington Weekly Free Press*, December 3, 1836, 3.
13. *Burlington Weekly Free Press*, December 30, 1836, 2.
14. *Burlington Weekly Free Press*, November 10, 1837, 3.
15. *Sentinel and Democrat* (Burlington, VT), January 7, 1839, 3.
16. *Burlington Weekly Free Press*, September 11, 1840, 4.
17. *Burlington Weekly Free Press*, October 11, 1839, 4.
18. Child, *Gazetteer*, 113.
19. Vermont Supreme Court, *Shaw v. Carpenter* (54 VT 155), 1872.

20. Hon. Hiram Carelton, *Genealogical and Family History of the State of Vermont* (New York: Lewis Publishing Co., 1903), 216–217.

21. *Burlington Weekly Free Press*, November 15, 1906, "Chronicling America: Historic American Newspapers," Library of Congress.

22. Accessed on March 14, 2014, https://www.sec.state.vt.us/media/59772/1847.pdf.

23. *St. Albans Daily Messenger*, February 3, 1903, 4.

24. Ibid.

25. Ibid.

26. Ibid.

27. Accessed on February 11, 2014, vermont-archives.org/govhistory/Referendum/pdf/1853.pdf.

28. *Revised State Laws of Vermont* (Rutland, VT: Tuttle & Co., 1881), 734.

29. *Vermont Watchman and State Journal* (Montpelier, VT), February 3, 1853, 2.

30. *Spirit Of The Age* (Woodstock, VT), January 10, 1903, "Chronicling America: Historic American Newspapers," Library of Congress.

1902

31. *Randolph Herald and News*, May 22, 1902, 2.

32. *Chelsea Herald*, May 22, 1902, 2.

33. *Randolph Herald and News*, May 22, 1902, 2.

34. *Bethel Currier*, May 22, 1902, 2.

35. Manuscript, author's collection.

36. Ibid.

37. Mason Green, *Nineteen-Two In Vermont: The Fight For Local Option Ten Years After* (Rutland, VT: Marble City Press, 1912), 16.

38. Green, *Nineteen-Two*, 21.

39. Green, *Nineteen-Two*, 154.

40. Green, *Nineteen-Two*, 34.

41. *New York Times*, August 31, 1902, 3.

42. https://www.sec.state.vt.us/media/48806/McCullough1902.pdf.

43. Joseph Battell, "The Present Crisis. Shall Private Gain be tolerated in the sale of intoxicating liquors? An appeal to Vermont Voters" (NP, c. 1902), 10.

44. Joseph N. Harris, *Prohibition, or Local Option?* (Ludlow, VT: 1902), 4. UVM Special Collections (HV5086.V5 H36 1902*)*.

45. The *Literary Digest*, February 14, 1903, vol. XXVI (New York: Funk & Wagnalls Company), 218.

46. Vermont State Archives, Middlesex, *Caledonia County Court*, Box 74, Folder 10, No. 2596, *State v. G. Ceruti*.

47. *The Local Option License and State-Wide Prohibition*, University of Vermont Bailey Howe Library, Special Collections, Wilb HV5086.VS T69 1908.

Interlude

48. Sally Pollak, "Maple Sugarer Aims for 'Best Spirits in the World,'" *Burlington Free Press*, April 3, 2015, retrieved August 11, 2018, www.burlingtonfreepress.com/story/news/local/2015/04/02/maple-spirits-franklin-county-farm/70837512/.

Vermont Pub & Brewery

49. Brewers Association, "History of Craft Brewing," retrieved July 28, 2018, www.brewersassociation.org/brewers-association/history/history-of-craft-brewing/.

50. Jeff S. Baker II, interview with Steve Polewacyk, October 13, 2017.

51. Michael Tonsmeire, "Book Review: New Brewing Lager Beer," *The Mad Fermentationist*, May 16, 2009, retrieved July 28, 2018, www.themadfermentationist.com/2009/05/book-review-new-brewing-lager-beers.html.

52. Janet Essman Franz, "Tap Dancing," *Business People-Vermont*, February 2007, retrieved July 28, 2018, businesspeoplevermont.com/2007/02-feb/pub.html.

53. Ibid.

54. Karen Kane, "Brewer's Profile: Greg Noonan," *Yankee Brew News*, January 1996.

55. Kurt Staudter and Adam Krakowski, *Vermont Beer: History of a Brewing Revolution* (Charleston, SC: The History Press, 2014), 66–67.

56. Personal correspondence with Steve Polewacyk, August 25, 2018.

57. Personal correspondence with Steve Polewacyk, August 5, 2018.

58. Jeff S. Baker II, interview with Steve Polewacyk, October 13, 2017.

59. Corin Hirsch, "A 25th Anniversary for Vermont Pub & Brewery," *Seven Days*, November 5, 2013, retrieved February 1, 2018, www.sevendaysvt.com/vermont/a-25th-anniversary-for-vermont-pub-and-brewery/Content?oid=2266521.

60. Ibid.

61. Rebecca Kirkman, "Sap on Tap: Brewing with Liquid Gold," *BeerAdvocate*, Issue No. 111, April 2016, 21.

62. Tim Camerato, "Salt Hill Pub Buys Seven Barrel," *Valley News*, November 5, 2016, retrieved July 28, 2018, www.vnews.com/Seven-Barrel-sold-to-Salt-Hill-Owners-6378802.

63. Phone correspondence with Steve Polewacyk, October 13, 2017.

64. Jeff S. Baker II and Jason Strempek, "Green Mountain Mashers," March 28, 2017, *It's the Beer Talking* (podcast), *Burlington Free Press*.

65. Noonan also wrote works of fiction, as well as history pieces on the origins of his surname and ancient Irish origin stories. More information can be found at http://www.nemhnain.com/.

66. Staudter and Krakowski, *Vermont Beer*, 94.

67. "Recognition Award," American Homebrewers Association, 2017, retrieved August 5, 2018, www.homebrewersassociation.org/membership/aha-governing-committee/recognition-award/.

68. "Greg Noonan," Brewers Publications, retrieved July 9, 2018, www.brewerspublications.com/blogs/author/greg-noonan; "Craft Brewing Industry Awards." Brewers Association, 2018, retrieved August 5, 2018, www.brewersassociation.org/press-room/awards/.

69. "Remembering Greg Noonan," Vermont Pub & Brewery, retrieved July 30, 2018, www.vermontbrewery.com/remebering-greg-noonan.

70. With apologies to the poet John Donne (JSB II).

71. Paul Kowalski, "What's Brewing in Vermont," *Yankee Brew News*, vol. 24, no. 03, June/July 2013, 33.

72. Sean Williams (n.d.), posts (LinkedIn Page), retrieved August 5, 2018, https://www.linkedin.com/in/sean-williams-51664391/.

73. Staudter and Krakowski, *Vermont Beer*, 87–88, 104.

74. Pip Vaughan-Hughes, "Beer Buzz," Business People: Vermont, August 1999, retrieved November 15, 2017, www.vermontguides.com/1999/8-aug/aug2.htm.

75. "Magic Hat, Vermont Pub & Brewery Brew up Steven Sour Sour IPA for Anniversary (Video)," edited by Adam Nason, BeerPulse, November 9, 2013, retrieved August 5, 2018, beerpulse.com/2013/11/magic-hat-vermont-pub-brewery-brew-up-steven-sour-sour-ipa-for-anniversary-video-1824/; in the promotional video referenced here, Polewacyk gives two alternative accounts of how they came up with the idea of the Steven Sour recipe. One includes a "giant eagle with a forty-foot wingspan" that snatched Polewacyk and took him up to the eagle's nest. Once there, the

eagle spoke to him in Latin and Mandarin Chinese, then smiled (which Polewacyk notes is difficult for an eagle) and told him to brew a sour IPA. The other version he offers is that maybe the brewers were just sitting around drinking beers—"one or the other."

76. Stan Hieronymus, "Greg Noonan in His Own Words," Appellation Beer, October 13, 2009, retrieved June 25, 2018, appellationbeer.com/blog/greg-noonan-in-his-own-words/.

77. Hirsch, "25th Anniversary."

78. Baker II, interview with Steve Polewacyk.

79. Steve Butcher, "Spruce Tip IPA," Lawson's Finest Liquids, March 12, 2014, retrieved July 29, 2018, www.lawsonsfinest.com/beers/spruce-tip-ipa/.

80. Shaun e. Hill, "A Quiet Return Home...," Hill Farmstead Brewery, November 23, 2009, retrieved August 8, 2017, hillfarmstead.com/main/2009/11/23/a-quiet-return-home.html.

81. "Green Mountain Mashers Announces Winners of Greg Noonan Memorial Homebrew Contest," VTDigger.org, Green Mountain Mashers, May 17, 2015, retrieved August 5, 2018, vtdigger.org/2015/05/17/green-mountain-mashers-announces-winners-of-greg-noonan-memorial-homebrew-contest/.

82. "The Greg Noonan New England Brewer's Scholarship," American Brewers Guild, 2018, retrieved July 28, 2018, www.abgbrew.com/index.php/admissions/scholarships/noonan.

The Black IPA

83. Jack Curtin, "The Name Game," Jack Curtin's Liquid Diet, January 13, 2011, jackcurtin.com/ldo/?p=3258.

84. Ibid.

85. Lauren Buzzeo, "Smuttynose Noonan Black IPA," Wine Enthusiast Magazine, September 1, 2013, www.winemag.com/buying-guide/smuttynose-noonan-black-ipa/.

86. Frank Faulkner, The Theory And Practice of Modern Brewing: A Re-written and much enlarged edition of "The Art of Brewing" (London: F. W. Lyon, 1888), 260.

Magic Hat Brewing Company

87. Alan Newman, *High on Business: The Life, Times, and Lessons of a Serial Entrepreneur* (Public Press, 2011), 47 and 77.

88. Ibid., 124–125.

89. Ibid., 76–77.

90. Ibid., 78.

91. Staudter and Krakowski, *Vermont Beer*, 77–78.

92. John Holl, "Magic Hat Brewing Company," *The Oxford Companion to Beer*, edited by Garrett Oliver (Oxford University Press, 2012), 557–558.

93. Jeff S. Baker II and Jason Strempek, "Alan Newman," June 6, 2017, *It's the Beer Talking* (podcast), *Burlington Free Press*.

94. Mark Hegedus, "Official History of the Magic Hat Brewing Company," sent via email, February 3, 2018.

95 Personal correspondence with Katie Dewitt of Magic Hat, February 14, 2018.

96. Hegedus, "Official History of Magic Hat."

97. Newman, *High on Business*, 161–162.

98. Although Magic Hat is now located in South Burlington, its origins in—and ultimate influence on—the Burlington beer scene cannot be understated nor omitted from this work.

99. Newman, *High on Business*, 147

100. Hegedus, "Official History of Magic Hat."

101. Dan Bolles, "From Phish Shows to Breweries, Builder Russ Bennett Makes Magic," *Seven Days*, September 28, 2016, www.sevendaysvt.com/vermont/from-phish-shows-to-breweries-builder-russ-bennett-makes-magic/Content?oid=3703769.

102. As a side note, Christopher Rockwood is one of many Vermont brewers who have earned a degree in philosophy. Others include Paul Sayler (Zero Gravity Craft Brewery) and Shaun e. Hill (Hill Farmstead); Brittney Hibbs-Kelson, director, "What's On Tap?: New Variety at Magic Hat," *My Champlain Valley*, June 4, 2018, www.mychamplainvalley.com/news/what-s-on-tap-new-variety-at-magic-hat/1217418006.

103. Lauren Ober, "Magic Hat Produces Its Own Energy—With Beer," *Seven Days*, January 26, 2011, www.sevendaysvt.com/vermont/magic-hat-produces-its-own-energy-with-beer/Content?oid=2142409.

104. Personal correspondence with Adam Fuller of Magic Hat, February 26, 2018.

105. Newman, *High on Business*, 171; Holl, "Magic Hat Brewing Company."

106. Newman, *High on Business*, 173.

107. Holl, "Magic Hat Brewing Company."

108 "KPS Capital Partners to Sell North American Breweries to Cerveceria Costa Rica, S.A.," KPS Capital Partners, LP, October 26, 2012, www.kpsfund.com/news/press-releases/2012/10/26/kps-capital-partners-to-sell-north-american-breweries-to-cerveceria-costa-rica-sa.

109. Hannah Palmer Egan, "Magic Hat Opens Full Tasting Room With Food, and a Party," *Seven Days*, May 3, 2017, www.sevendaysvt.com/vermont/magic-hat-opens-full-tasting-room-with-food-and-a-party/Content?oid=5474050.

110. Sadie Williams, "Magic Hat Receives National Arts Award," *Seven Days*, May 11, 2017, www.sevendaysvt.com/LiveCulture/archives/2017/05/11/magic-hat-receives-national-arts-award; "BCA 10," Americans for the Arts, 2018, www.americansforthearts.org/events/bca-10.

111. Liza Gershman, *Drink Vermont: Beer, Wine, and Spirits of the Green Mountain State* (Skyhorse Publishing, 2017), 79.

112. Newman, *High on Business*, 79.

113. Jeff S. Baker II and Adam Krakowski, interview with Paul Sayler, April 19, 2017.

114. "Magic Hat to Release Feast of Fools Holiday Stout at Fundraiser," Brewbound.com, November 16, 2017, www.brewbound.com/news/magic-hat-release-feast-fools-holiday-stout-fundraiser.

115. Andy Crouch, "The Freedom to Brew," BeerScribe.com, 2003, www.beerscribe.com/freedom.html.

116. Vermont Act 167, 2008, AN ACT RELATING TO RETAIL SALES AND TAXING OF SPECIALTY BEERS.

117. Michael Jackson, et al., "Review: Heart of Darkness Stout," *All About Beer*, March 1, 2001, allaboutbeer.com/review/heart-of-darkness-stout/.

118. Todd Alström, "Review: Ale Of The Living Dead," *BeerAdvocate*, October 17, 2001, www.beeradvocate.com/beer/profile/96/1775/?ba=bros.

119. Hannah Palmer Egan, "Talking Shop With Fiddlehead Brewer Matt Cohen," *Seven Days*, November 14, 2014, www.sevendaysvt.com/vermont/talking-shop-with-fiddlehead-brewer-matt-cohen/Content?oid=2475082.

120. "Big Hundo," Christopher Rockwood and Heather Darby, YouTube, Magic Hat Brewing Company, June 15, 2017, youtu.be/qp_2HdDQz9I.

121. "Alchemy & Science Joins Forces with The Boston Beer Company," Alchemy & Science, October 28, 2011, http://asbeer.com/alchemy-science-joins-forces-with-the-boston-beer-company/.
122. Baker II and Strempek, "Alan Newman."

Three Needs Taproom & Brewery

123. Jeff S. Baker II, interview with Glenn Walter, July 6, 2017.
124. Bettina Cataldi, "Burlington's Hierarchy of Needs," *The Vermont Cynic*, February 28, 2018, vtcynic.com/features/burlingtons-hierarchy-of-needs/.
125. Baker II, interview with Glenn Walter.
126. Dan Kochakian, "Contract Brewing at Ipswich Ale Brewery," *Ale Street News*, vol. 22, issue 4, August/September 2013, 4B.
127. Pip Vaughan-Hughes, "Beer Buzz." Business People: Vermont, August 1999, retrieved November 15, 2017, www.vermontguides.com/1999/8-aug/aug2.htm.
128. Kochakian, "Contract Brewing at Ipswich."
129. Jeffrey Hughes, (n.d.) posts (LinkedIn page), retrieved July 14, 2018, https://www.linkedin.com/in/jeffrey-hughes-23127a28/.
130. Andy Crouch, *The Good Beer Guide to New England* (Lebanon, NH: University Press of New England, 2006), 235–237.
131. James Tierney, (n.d.) posts (LinkedIn page), retrieved July 14, 2018, https://www.linkedin.com/in/james-tierney-86a9432a/.
132. Personal correspondence with Scott Martin via email, December 1, 2017.
133. William S.A. Floyd, "Pearl Street East from Corner of Church Street," Church Street Top Blocks, Burlington, Vermont, University of Vermont, December 7, 2017, www.uvm.edu/~hp206/2017/pages/Floyd/index.html.
134. Simon Davidson, "Bodo's," *The Charlottesville*, June 29, 2012, accessed July 14, 2018, charlottesville29.com/bodos/.
135. William Braun, "The Déjà Vu," *Burlington Free Press*, November 27, 1977, 6–8.
136. Eloise Hedbor, "Restaurateur Found Dead in Home," *Burlington Free Press*, October 5, 1985, 2B; Page, Candace, "Déjà Vu Changes Owners," *Burlington Free Press*, December 30, 1986, 6A.

137. "Hot Properties: Secrets of the City," *Metropolitan Home*, November 1987, 34.

138. Paula Routly, "A Fuller Plate: Managing Multiple Menus Keeps One Restaurateur on the Run," *Seven Days*, May 15, 2002, www.sevendaysvt.com/vermont/a-fuller-plate-managing-multiple-menus-keeps-one-restaurateur-on-the-run/Content?oid=2552213.

139. Hannah Palmer Egan, "Three Needs Resumes Brewing," *Seven Days*, May 20, 2015, www.sevendaysvt.com/vermont/three-needs-resumes-brewing/Content?oid=2612497.

140. Baker II, interview with Glenn Walter.

Switchback Brewing Company

141. Chris McClellan, "Switchback Brewing Company," The Brew Enthusiast, September 2015, retrieved August 11, 2018, www.thebrewenthusiast.com/switchback-brewing-company/.

142. "Meet The Crew," Switchback Brewing Company, switchbackvt.com/meet-the-crew.html.

143. "Co-Owner Clips: Tony Morse," Switchback Brewing Company, switchbackvt.com/co-owner-clips-tony-morse.html.

144. Personal correspondence with Todd Haire, November 16, 2017.

145. Jeff S. Baker II, interview with Billy Cherry, December 15, 2017.

146. Jeff S. Baker II, "Brewer to Brewer: Fiddlehead & Switchback," The Daily Beet, September 11, 2015, retrieved March 28, 2017, read.localvore.com/brewer-to-brewer-fiddlehead-switchback/.

American Flatbread Burlington Hearth

147. "Brewpub at Flatbread," Zero Gravity Craft Brewery, retrieved July 21, 2018, www.zerogravitybeer.com/brewpub-at-flatbread/.

148. Adam Krakowski and Jeff S. Baker II, interview with Paul Sayler, April 19, 2017.

149. "115 St. Paul Street (Now 115–119 St. Paul Street)," Historic Burlington, Vermont, University of Vermont, December 7, 2004, retrieved July 22, 2018, www.uvm.edu/~hp206/2004-1890/burlington1890/website/sprull/Addresses/?Page=115st.paul.html.

150. Alice Levitt, "Deceased Feast," *Seven Days*, October 31, 2012, www. sevendaysvt.com/vermont/deceased-feast/Content?oid=2241966.

151. Personal correspondence with Paul Limberty, July 14, 2018.

152. Hedbor, "Restaurateur Found Dead in Home."

153. "Carbur's Shooting Apparent Suicide," *Burlington Free Press*, January 10, 1992, 3B.

154. Thea Lewis, *Haunted Burlington: Spirits of Vermont's Queen City* (Charleston, SC: The History Press, 2009), 77–82.

155. Ibid. Local medium Nan O'Brien proposed that this building was part of a smuggling network between Canada and the United States in the late 1700s and early 1800s and that the spirits are those of marauding and scandalous pirates.

156. Krakowski and Baker II, interview with Paul Sayler.

157. Jackie Daub, "Let's Talk Chocolate with Zero Gravity Brewery," Lake Champlain Chocolates, August 9, 2016, retrieved July 22, 2018, www.lakechamplainchocolates.com/blog/lets-talk-chocolate-with-zero-gravity-brewery/.

158. Hannah Palmer Egan, "On Tap: Zero Gravity Craft Brewery," *Seven Days*, January 7, 2015, www.sevendaysvt.com/vermont/on-tap-zero-gravity-craft-brewery/Content?oid=2499422.

159. Chas. W. Parsons, ed., "Special Announcements for the Drug Trade," *The Pharmaceutical Era*, November 15, 1894, 42; Staudter and Krakowski, *Vermont Beer*, 118.

160. Ellyn Gados, "Tonic Inebriation," *Edible Green Mountains*, August 15, 2015, retrieved July 22, 2018, ediblegreenmountains.ediblecommunities.com/drink/tonic-inebriation.

161. Jeremy Egner, "36 Hours in Burlington, Vermont," *New York Times*, August 19, 2015, retrieved June 15, 2018, www.nytimes.com/2015/08/23/travel/what-to-do-in-36-hours-in-burlington-vermont.html; Todd Plummer, "A Craft Beer Crawl Through Burlington, Vermont," *Vogue*, May 9, 2017, retrieved July 22, 2018, www.vogue.com/article/burlington-vermont-craft-beer-crawl.

162. "Daysies Finalists 2018," Seven Days, retrieved July 22, 2018, https://posting.sevendaysvt.com/vermont/DaysiesFinalists2018/Page.

Zero Gravity Craft Brewery

163. "Zero Gravity Brewing Mural, Burlington VT," Fontana Studios, October 2015, retrieved August 11, 2018, thefontanastudios.com/mural/zero-gravity-brewing-mural.
164. Staff, "Harpoon Brewery Releases 100 Barrel Series Braggot Rights," All About Beer, March 25, 2015, retrieved August 11, 2018, allaboutbeer.com/news/harpoon-brewery-releases-100-barrel-series-braggot-rights/.

Queen City Brewery

165. Jeff S. Baker II and Jason Strempek, "Episode 006: Queen City Brewery," August 9, 2016, *It's the Beer Talking* (podcast), *Burlington Free Press*.
166. Pamela Hunt, "Queen City Brewery Goes Old-School to Offer Vermont Steinbier," CraftBeer.com, Brewers Association, March 27, 2017, www.craftbeer.com/featured-brewery/queen-city-brewery-vermont-steinbier.
167. Retrieved March 26, 2017, https://queencitybrewery.wordpress.com/2012/06/24/hot-rocks-steinbier-6-2/.
168. Kavet, Rockler, & Associates, LLC., "The Economic Impact of Craft Brewing on the Vermont Economy in 2014: Prepared for the Vermont Brewers Association," November 2015, 5.
169. Jeff S. Baker II, "Brewer to Brewer: A Tale of Two Pauls in the Queen City," The Beet, July 01, 2015, http://read.localvoretoday.com/brewer-to-brewer-a-tale-of-two-pauls-in-the-queen-city/.
170. Classic Beer Style No. 8: Scotch Ale, Gregory J. Noonan, Brewers Publications, 1998.
171. Retrieved March 26, 2017, https://queencitybrewery.wordpress.com/2016/07/12/queen-city-announces-new-head-brewer/.
172. *Historic Guide to Burlington Neighborhoods*, vol. 2, Queen City Printers (Chittenden County Historical Society), 1997.

Simple Roots Brewing

173. Jeff S. Baker II, Simple Roots interview, June 05, 2017.
174. Via email June 27, 2017.

175. Joshua M. Bernstein, "Road Tripping: Exploring 10 of America's Best Below-the-Radar Beer Destinations," *Imbibe Magazine*, September/October 2017, 42.

176. Originally called Vermont Butter & Cheese Company, Vermont Creamery was founded in 1984 in Brookfield, Vermont, by Allison Hooper and Bob Reese. The creamery moved to Websterville, Vermont, in 1989, and in 2017, Hooper and Reese announced that they were selling the company to Land O'Lakes, a farmer-owned co-operative headquartered in Arden Hills, Minnesota.

Foam Brewers

177. Todd Haire, "An Ode to Foam," Foam Brewers, 2016, www.foambrewers.com/.

178. Gregory S. Jacobs, "Mills & Factories: Manufacturing Heritage Sites in Burlington and Winooski, Vermont," UVM Historic Preservation Program, University of Vermont, March 1, 2014, www.uvm.edu/~hp206/2013/pages/jacobs/index.html.

179. Ibid.

180. Jeff S. Baker II, Foam Brewers interview, May 14, 2017.

181. Gershman, *Drink Vermont*, 66.

182. Jeff S. Baker II and Jason Strempek, "Foam Brewers & House of Fermentology," March 7, 2017, *It's the Beer Talking* (podcast), *Burlington Free Press*.

183. Gershman, *Drink Vermont*, 66.

184. Julia Clancy, "Burlington's New Foam Brewers Go With the Flow," *Seven Days*, September 14, 2016, www.sevendaysvt.com/vermont/burlingtons-new-foam-brewers-go-with-the-flow/Content?oid=3667510.

185. Baker II, Foam Brewers interview.

186. Baker II and Strempek, "Foam Brewers & House of Fermentology."

187. Clancy, "Burlington's New Foam Brewers."

188. Mike Sardina, "Great Beer Coming to a Head—Foam Brewers in Burlington, Vermont," Good Beer Hunting, January 17, 2018, www.goodbeerhunting.com/blog/2018/1/15/great-beer-coming-to-a-head-foam-brewers-of-burlington-vermont.

189. Baker II, Foam Brewers interview.

190. The "Iceman pour," a.k.a. the "boss pour," refers to a trend that began around 2015. Drinkers would pour a beer into a glass right up to the

top without any foam at all and post a photo of it to social media. Many brewers believe this is disrespectful because it doesn't allow for proper enjoyment of the aromatics. See "Surface Tension: The Probably Untrue Story of the Iceman Pour," GoodBeerHunting.com.

191. Jeff S. Baker II, "Where Is the Craft Beer Market Heading?" *Burlington Free Press*, February 17, 2017, 4C.

192. Baker II and Strempek, "Foam Brewers & House of Fermentology."

193. Clancy, "Burlington's New Foam Brewers."

194. "Class of 2016: 34 of the Best New Breweries in the US," *BeerAdvocate*, no. 120, January 2017, 49; "Best New Brewers in the World," RateBeer.com, January 30, 2017, www.ratebeer.com/ratebeerbest/newbrewers_2017.asp.

195. "LCRCC Recognizes Foam Brewers with Entrepreneurial Spirit Award," Lake Champlain Regional Chamber of Commerce Blog, November 9, 2017, www.vermont.org/chamber/blog/2017/11/09/lcrcc-recognizes-foam-brewers-with-entrepreneurial-spirit-award/.

196. "Daysies Finalists 2018," *Seven Days*.

Vermont: Land of IPAs

197. Ethan Fixell, "How Vermont Became the New IPA King," *Men's Journal*, December 4, 2017, www.mensjournal.com/food-drink/how-vermont-became-the-new-ipa-king-w203293/.

198. Staudter and Krakowski, *Vermont Beer*, 75.

199. Jeffrey S. Baker, "Hops & Barley: Easy to Make a Case for the Vermont IPA," *Burlington Free Press*, March 16, 2012, 2D.

200. Stan Hieronymous, "Gather No Moss: The Perpetual Evolution of Heady Topper," *Craft Beer & Brewing Magazine*, April 2014, 82–83.

201. Personal correspondence with Conor Giard via email, October 8, 2017.

202. Hill Farmstead Brewery, "Hill Farmstead Brewery Named Best Brewery in the World," accessed April 20, 2019, http://hillfarmstead.com/main/2019/1/29/hill-farmstead-named-best-brewery-in-the-world-for-2018.html.

203. Personal correspondence with Sam King via email, November 12, 2017.

204. Baker, "Hops & Barley."

205. Ken Weaver, "The Case of the Hazy IPA," *All About Beer Magazine*, January 2017, allaboutbeer.com/article/hazy-ipa/; Michael Kiser, "Critical Drinking with The Alchemist's John Kimmich," Good

Beer Hunting, October 11, 2016, www.goodbeerhunting.com/blog/2016/10/4/critical-drinking-with-john-kimmich-of-the-alchemist.

206. Hill, "A Quiet Return Home..."

207. Personal correspondence with Shaun e. Hill via text message, October 7, 2017.

208. The beer's name comes from a nearby unfinished road project called the "Southern Connector." The tap-handle design for Connector IPA is in the shape of a tall filing cabinet, a reference to "the world's tallest filing cabinet," an art installation on Flynn Avenue made by local artist Bren Alvarez to represent the bureaucracy of the construction project.

209. Brad Japhe, "Vermont-Style IPA: A Beer That's a Little Bit Juicy," Eater.com, November 5, 2015, www.eater.com/drinks/2015/11/5/9676482/vermont-style-ipa-alchemist.

210. "Editors' Picks: Imperial Pale Ales," *Craft Beer & Brewing Magazine*, February/March 2015, 86.

211. Hannah Palmer Egan, "On Tap: A Conversation With Alchemist Brewer John Kimmich," *Seven Days*, May 6, 2015, www.sevendaysvt.com/vermont/on-tap-a-conversation-with-alchemist-brewer-john-kimmich/Content?oid=2583108.

212. Joshua M. Bernstein, "Eastern Promise," *BeerAdvocate*, issue no. 93, October 2014, 56.

213. Jeff S. Baker II, "Flattery or Unfair? 'Vermont' Beer Pops up out of State," *Burlington Free Press*, March 17, 2017, 1C–3C.

214. Zach Fowle, "An American Beer Style in London," *DRAFT Magazine*, September 2, 2016, http://draftmag.com/an-american-beer-style-in-london/.

215. It should be noted that in 2014, Harpoon Brewery used the phrase "definitive New England-style IPA" in marketing materials for its flagship IPA but has since dropped this description.

216. Jason and Todd Alström, "It's Official: New England India Pale Ale Is a Style," *BeerAdvocate*, May 2017, 2.

217. Personal correspondence with Chris Swersey via phone, June 16, 2018.

House of Fermentology

218. Bill Mares and Todd Haire, *Making Beer: From Homebrewing to the House of Fermentology* (Shelburne, VT: Bard Owl Books, 2016), 116.

219. Ibid.

220. Jeff S. Baker II and Jason Strempek, "Foam Brewers & House of Fermentology," March 7, 2017, *It's the Beer Talking* (podcast), *Burlington Free Press*.
221. Ibid.
222. It is the opinion of the authors that both Haire and Mares look dashing with said mustaches.
223. Mares and Haire, *Making Beer*, 114.
224. Ibid., 95.
225. Baker II and Strempek, "Foam Brewers & House of Fermentology."
226. Mares and Haire, *Making Beer*, 115–117.
227. Ibid., 115.
228. Jeff S. Baker II, Foam Brewers interview, May 14, 2017.
229. Mares and Haire, *Making Beer*, 121.
230. Ibid., 119.

Taprooms

231. Jeff S. Baker II, "What Is a Taproom?" *Burlington Free Press*, June 15, 2012, 9C.
232. Jeff S. Baker II and Adam Krakowski, interview with Paul Sayler, April 19, 2017.
233. Personal correspondence with Jed Davis via email, October 11, 2017.
234. "Fall Beers: They Speaketh and We Drinketh," *Stowe Today*, December 31, 2014, www.stowetoday.com/this_week/food/fall-beers-they-speaketh-and-we-drinketh/article_f9c5dc26-0618-11e1-bf15-001cc4c002e0.html.
235. Alice Levitt, "Parlor Trick," *Seven Days*, December 22, 2010, www.sevendaysvt.com/vermont/parlor-trick/Content?oid=2142182.
236. Suzanne Podhaizer, "Gourmet Onion," *Seven Days*, October 10, 2006, www.sevendaysvt.com/vermont/gourmet-onion/Content?oid=2131947.
237. Personal correspondence with Justin Gould via email, January 22, 2018.
238. Jeff S. Baker II and Jason Strempek, "Episode 02.06," February 28, 2017, *It's the Beer Talking* (podcast), *Burlington Free Press*.
239. Mike Dunphy, "Beer Destinations: Burlington, Vermont," *BeerAdvocate*, September 2013, 44; Austin L. Ray, "Barkeep: The Farmhouse Tap & Grill," *BeerAdvocate*, November 2014, 6; Heather Vandenengel, "The New Burlington Beer Scene," *Craft Beer & Brewing Magazine*, Summer 2014, 28; Allen McDuffee, "Beertown, U.S.A.: Burlington, Vt.," *DRAFT Magazine*, March 25, 2015, http://draftmag.com/beertown-u-s-a-burlington-vt/.

240. Lewis, *Haunted Burlington*, 53–56.
241. McDuffee, "Beertown, U.S.A."

Pint Street

242. Joshua M. Bernstein, "Road Tripping: Exploring 10 of America's Best Below-the-Radar Beer Destinations," *Imbibe Magazine*, September/October 2017, 42.
243. Sally Pollak, "Pine Street: From Wasteland to Tasteland," *Burlington Free Press*, August 29, 2014. http://www.burlingtonfreepress.com/story/life/2014/08/29/pine-street-foodie-destination/14754699/.
244. Paul Hale, "Pine Street Is Now Pint Street," Queen City Brewery, June 10, 2015, https://queencitybrewery.wordpress.com/2015/06/10/pine-street-is-now-pint-street/; Paul Hale (@QueenCtyBrewery), Tweet, "Pine Street is now Pint Street!" June 10, 2015, 12:36 p.m., https://twitter.com/QueenCtyBrewery/status/608659066357972992.
245. Sally Pollak, "The Great Northern Opens in Burlington," *Seven Days*, May 16, 2017, https://www.sevendaysvt.com/vermont/the-great-northern-opens-in-burlington/Content?oid=5708990.
246. Sally Pollak, "The Great Northern's Vintage Backbar Is Storied," *Seven Days*, June 21, 2017, https://www.sevendaysvt.com/vermont/the-great-northerns-vintage-backbar-is-storied/Content?oid=6338259.
247. Ibid.

Queen (City) of Tourism

248. Personal correspondence with Brittney Hibbs-Kelson, November 13, 2017.
249. Personal correspondence with Steven Cook via email, June 19, 2018.
250. Heather Vandenengel, "The New Burlington Beer Scene," *Craft Beer & Brewing Magazine*, 2014, 24–28; Dunphy, Mike, "Beer Destinations: Burlington, VT," *BeerAdvocate*, issue no. 80, September 2013, 42–44; Bernstein, "Road Tripping"; Jeremy Egner, "36 Hours in Burlington, Vermont," *New York Times*, August 19, 2015, www.nytimes.com/2015/08/23/travel/what-to-do-in-36-hours-in-burlington-vermont.html; Lauren Duffy Lastowka, "Why Burlington Is the Best

Beer City in the USA," Serious Eats, June 2016, www.seriouseats. com/2016/06/burlington-best-beer-city.html.

251. Personal correspondence with Tommy Noonan.

252. Chelsea Davis, "The Country's First Beer Concierge Talks Brew and Business," *Paste Magazine*, May 2, 2017, www.pastemagazine. com/articles/2017/05/the-countrys-first-beer-concierge-talks-brew-and-b.html.

253. Jeff S. Baker II and Jason Strempek, "Matt Canning, Beer Concierge," October 25, 2016, *It's the Beer Talking* (podcast), *Burlington Free Press*.

254. "Summer Beer Exploration," Hotel Vermont, 2018, hotelvt.com/stay/ packages/winter-beer-tour-package/.

255. "Vermont Brewers Festival Burlington," Vermont Brewers Association, www.vermontbrewers.com/festivals/vermont-brewers-festival-burlington/.

256. Sally Pollak, "Vermont Brewers Festival Sold Out in 11 Minutes," *Burlington Free Press*, May 16, 2014, retrieved May 12, 2018, www.burlingtonfreepress. com/story/life/2014/05/16/vermont-brewers-festival-sold/9184063/.

257. "Best Bartender," *Seven Days* Daysies Awards, 2015 and 2016.

258. https://www.citybrewtours.com/burlington/.

259. Jeff S. Baker II and Jason Strempek, "The Business of Craft Beer," July 9, 2017, *It's the Beer Talking* (podcast), *Burlington Free Press*.

260. Personal correspondence with Brittney Hibbs-Kelson, November 13, 2017.

261. Joel Banner Baird, "Beer Podcast's Swan Song: VT Vets Shine at 14th Star Brewing," *Burlington Free Press*, June 19, 2018, www.burlingtonfreepress. com/story/life/2018/06/19/podcast-vermont-vets-shine-14th-star-brewing/661517002/.

262. Staudter and Krakowski, *Vermont Beer*, 99.

INDEX

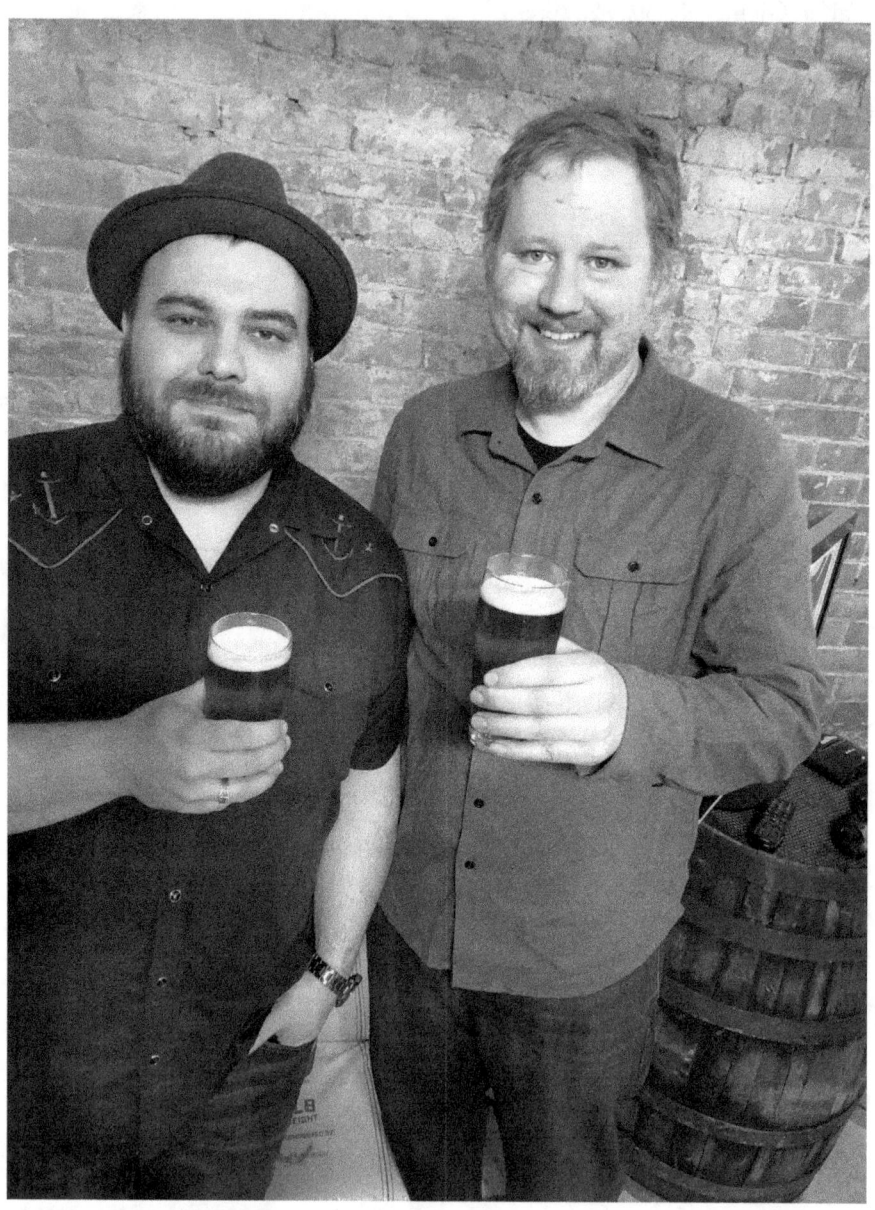

Jeff S. Baker II (*left*) and Adam Krakowski. *Courtesy of the authors.*

ABOUT THE AUTHORS

Adam Krakowski is a decorative and fine arts conservator who has worked at museums, historical societies, art galleries and restoration firms all over New York and New England. Adam coauthored *Vermont Beer: History of A Brewing Revolution* (The History Press, 2014), and authored *Vermont Prohibition: Teetotalers, Bootleggers, Corruption* (The History Press, 2016). He writes for *Yankee Brewing News*, a brewing industry newspaper, and contributes to other regional and national publications. In 2010, he was the recipient of the Weston Cate Jr. Research Fellowship from the Vermont Historical Society on the project, "A Bitter Past: Hop Farming in Nineteenth-Century Vermont." His focus of interest in his lecturing and writing is the beverage and food heritage of New England.

Jeff S. Baker II is a Certified Cicerone® and has worked for over a decade in the craft beer and fine wine industry in Florida, Massachusetts and Vermont. His fascination with the world of craft beverages began while earning a bachelor's degree in philosophy at St. Michael's College—*in vino veritas!*—and he has since earned multiple wine certificates. He curated an award-winning beer program for a Burlington farm-to-table restaurant and is now head of education and training for Farrell Distributing. A regular contributor to the *Burlington Free Press* and cohost of the *It's the Beer Talking* podcast, he has been nominated twice for "Best Print/Online Journalist in Vermont." In 2006, he received the Frank G. Mahady Memorial Award for Peace & Social Justice, and he is an ordained colonel. He lives with his partner, Jana Simmons, and their lovely dog, Harper, in the Burlington area.

Visit us at
www.historypress.com